TH 5667 .B58 1996

Blanc, Alan.

Stairs, Steps and Ramps

NEW ENGLAND INSTITUTE
OF TECHNOLOGY

STAIRS, STEPS AND RAMPS

Alan Blanc

Architectural Press
An imprint of Butterworth-Heinemann
Linacre House, Jordan Hill, Oxford OX2 8DP
A division of Reed Educational and Professional Publishing Ltd

R   A member of the Reed Elsevier plc group

OXFORD   JOHANNESBURG   BOSTON
NEW DELHI   SINGAPORE   MELBOURNE

First published 1996
Reprinted 1997

©Reed Educational and Professional Publishing Ltd 1996

All rights reserved. No part of this publication
may be reproduced in any material form (including
photocopying or storing in any medium by electronic
means and whether or not transiently or incidentally
to some other use of this publication) without the
written permission of the copyright holder except
in accordance with the provisions of the Copyright,
Designs and Patents Act 1988 or under the terms of a
licence issued by the Copyright Licensing Agency Ltd,
90 Tottenham Court Road, London, England W1P 9HE.
Applications for the copyright holder's written permission
to reproduce any part of this publication should be addressed
to the publishers

British Library Cataloguing in Publication Data
Blanc, Alan
   Stairs, Steps and Ramps
   I. Title
   721.83

ISBN 0 7506 1526 5

Library of Congress Cataloguing in Publication Data
Blanc, Alan
   Stairs, steps, and ramps/Alan Blanc.
   p.   cm.
   Includes bibliographical references and index.
   ISBN 0 7506 1526 5 (paper)
   1. Staircase.   2. Ramps (Walkways).   I. title.
   TH5667.B58   1995   94–38848
   694′.6–dc20   CIP

Typeset by Keyword Typesetting Services Ltd,
Wallington, Surrey
Printed in Great Britain

Contents

Foreword vii

1 Introduction 1

2 Historical review 7

3 Domestic stairs 37

4 Commercial stairs 68

5 Civic and public stairs 98

6 External stairs 130

7 Detailed construction: timber 159

8 Detailed construction: iron, steel and other metals 181

9 Concrete stairs 209

10 Stonework and composite construction 228

11 Design codes and procedures 236

12 Elevators and mechanical circulation 259

13 International case studies 291

14 Case studies in the UK 317

Index 349

Foreword

In his later years Alan Blanc was a profilic writer and it is by good fortune that he completed *Stairs, Steps and Ramps* although a great shame that he didn't live to see it in print. Alan with his wife Sylvia had travelled prodigiously in the pursuit of architecture, as this book's illustrations bear witness, and its subject reflects one of their particular enthusiasms.

The organization of a pattern of circulation is a principle ordering device in contemporary buildings. The experience of an unfolding spatial sequence found in many of the best of modern architecture has much to do with the the elaboration of stairs, steps and ramps. Although the multitude of types from the humble and domestic to the grand and celebratory are recorded here, it is the element of repetition of tread upon tread, the way those repeated steps establish a rhythm of movement, an almost musical notation for a building that explains the continuing fascination with their design.

It is possible to read this book for the enjoyment of architectural episodes that stretch back to the beginnings of time, over the course of which the staircase and ramp have been invested with the greatest significance, for instance, the profound rituals represented by the great stepped structures of Asia Minor and Central America. It is also, however, a practical design guide with advice on all aspects from the design of individual components, to the legislative complications that impact on the planning of buildings as a whole. Included are today's mechanical equivalents, escalators, lifts and travelators to complete this unique and encyclopaedic view of the subject.

Alan Blanc was not only an architect who spent many years in practice but also a formidable teacher. His early experience of working with Walter Segal was a formative influence and through his teaching he ably communicated their shared interest in the craft of architecture. Alan loved the theatricality of giving a lecture, it was his great ebullience and enthusiasm that made its mark on a generation of students, and which is also to be found in these pages.

Mike McEvoy

1 Introduction

My life long obsession with stairs can be traced back to the London Blitz of 1940 while taking shelter night after night in the 'understairs' cupboard of a Victorian terraced house. The actual space was of triangular section, panelled on one side, with a rough brick party wall on the other, whilst overhead there existed the full panoply of the carpenter's art with glued and wedged treads and risers set into pine strings. The local ARP Warden advised 'under stairs' for sheltering as the strongest structure within the home when there was no official form of garden bunker or Morrison shelter. Being 'bombed out', and what to do next, were very much day-to-day topics, heightened by the bombing of two houses in the same street. To prepare for the event, young Blanc measured the family home and set about drawing up a design for the rebuilding – to accord with parental wishes. To guide the budding architectural ambition there were fat New Year copies of *Architect and Building News* stored conveniently under the stairs from the vintage years 1932, 1933 and 1934. The writer's discovery of staircases therefore began with trying to replan a slice of Victoriana, supposedly to be bombed but rebuilt for a better New World. That world was to be inspired in due course by further perusal of those dog-eared *Architect and Building News* with the works of Joseph Emberton, Edward Maufe and Owen Williams! It is little wonder that architecture became the vocation five years later.

The second turning point follows the sudden flush of funds that permitted a buying spree in Tiranti's *circa* 1949. The purchases included two volumes of *L'Architecture Vivante, en Allemagne* (Erich Mendlesohn & Weissenhof) together with my most prized aquisition Franz Schuster's *Treppen*. The book is pre-war but was resurrected and reprinted by Hoffmann of Stuttgart in 1949. The old Nazi style ingredients had not been eradicated so that the graceful designs of Aalto, Asplund or Salvisberg sit side by side with Hitlerian Youth Hostels and air raid shelters 'Versuch Sanstalt Fur Luftfahrt'. Schuster's finer hand is seen in the detailed advice with inspirational concern for geometry that makes the art of stair construction equal to furniture in terms of exactness and line (Figure 1.1). I have collected other books on stairs but have to confess that Schuster's *Treppen* is still my favourite.

The first essay in detailing came whilst working for Bertram Carter in 1949. Carter had just completed work for Dunn's of Bromley and through the same progressive client came the chance to build a concrete framed house with a stair using cantilever treads. The next step forward followed tutelage under Walter Segal when the pros and cons of Franz Schuster were fully aired, Segal having known his work pre-war. Working with

2 Stairs, Steps and Ramps

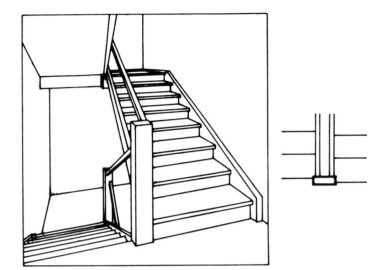

Figure 1.1 Variations at well end. (from Schuster, F., Treppen, Hoffman Verlag, 1949)

Figure 1.1a Solid newel at well end

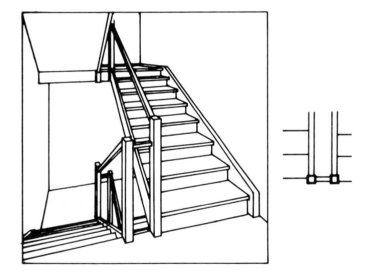

Figure 1.1b Twin newels at well end

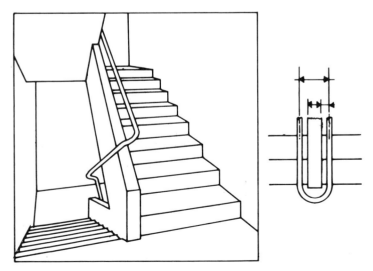

Figure 1.1c Swept handrail (balustrade not shown but will infill between rail and string)

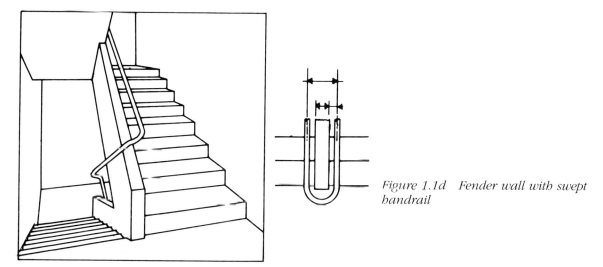

Figure 1.1d   Fender wall with swept handrail

Walter meant drawing out staircases to full size including the balustrade, pinning the tracing to the wall where the stair was to be installed in order that the visual effect could be fully established. In one case the balusters were whittled down by grooving to around 10 mm. The owner's daughter fell through the lower most balustrade and Walter wrote a cheque by return of post for bronze balusters throughout. Codes of Practice and Section K of the National Building Regulations rule out those risks today but it is worth recalling that the minimal elegance achieved in the designs of Eva Jiricna can only be obtained by working in close collaboration with a qualified engineer.

The historical references within *Treppen* started my interest in photographing stairs and in discovering the architecture of Gunnar Asplund, Roland Rohn or Otto Salvisberg that were drawn upon by Schuster. It is not possible to isolate the detailing solution from the context of the design. A typical example is Asplund's Courthouse at Gothenburg where the movement sequence from entry court to loggia and stairhall leads with formal splendour to the courtrooms at first floor and thence, by secondary means, to the offices at upper levels. Such a 'promenade in space' is perhaps the finest achievement that well designed stairs can achieve. The open vertical circulation eloquently arranged at the core of buildings can certainly become the most captivating experience. Designs as diverse as Garnier's Paris Opera (Figure 1.2), Le Corbusier's Villa La Roche-Jeanneret and Lasdun's National Theatre in London all reveal the magic that is possible with creative designers.

Today lifts and stairs as well as escalators form the core of all multi-storey plans, often sadly hidden behind solid enclosures to form uninteresting fire compartment zones. Modern technology using heat resistant glazing releases the architect from boxed-in solutions. Visual connection is once more feasible between vertical circulation and the spaces served. Pressure controlled environments can exclude smoke without recourse to the complication of double lobbies. Research for this book amongst older masterpieces in London often revealed that double lobbies are thwarted by having one set of doors permanently wedged open or simply propped ajar with a fire extinguisher.

Outstanding developments such as Lloyds HQ, London, or the Hongkong Shanghai Bank explore the new technologies and reveal that staircases and their modern equivalents can recapture the visual excitement that Piranesi (Figure 1.3) or Leonardo da Vinci awarded to stairs within the interior volume of their designs.

This new text book dedicated to stairs may well rid the writer of an obsession. The urgent intent is to impart others with enthusiasm until my supply of pictures, sketches and sundry anecdotes are exhausted.

a

b

Figure 1.2 Opera House, Paris, 1862-75. (Charles Garnier)
a Cut-away model of structure   b Entry to first flight

Stairs, Steps and Ramps 5

Figure 1.2c Half landing and balconies

6 Stairs, Steps and Ramps

Figure 1.3 An obsession with stairs, extract from Piranesi's Carceri, Plate VII. (From Scott, J., Piranesi, *Academy Editions, 1975)*

2 Historical review

It is not possible to write a complete history of stairs within one chapter. The intention here is to highlight the main forms that have developed and to help to explain the whys and wherefores of the geometry involved

2.1 Philosophical roots

It is necessary to underline the philosophical roots of staircases, particularly where the analogy of ascent with fear and reverence are associated. The archetypal pyramid forms of Assyria and Mexico making interesting comparisions.

The Chaldean ramped temples (Figure 2.1*a*) emulate the description of the path to Heaven given for the Tower of Babel in James edition, one finds a more poetic vision 'Go to, let us build a city and a tower, whose top may REACH unto heaven'. The idea inspired many painters (Figure 2.1*b*).

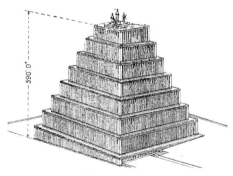

Figure 2.1 Philosophical roots
a Chaldean ramped temple. (From Fletcher, B., A History of Architecture, Batsford, 1945)

Genesis 11:4. There is an excellent reference to the materials used 'Come, let us make bricks, and burn them thoroughly'; further 'they had brick for stone, and bitumen for mortar'. Turning to the King

Figure 2.1b Detail from Bruegal The Tower of Babel

The ramped tower of the Assyrians with a winding route externally appears to be the inspiration for the early minarets (Figure 2.1*c*). The most sophisticated construction exists in Isfahan, where the stone treads are taken through the conical brickwork on both sides to form the ascending steps externally and the descending ones internally. The ascent is made into a gesture to be publicly celebrated.

Figure 2.1c Spiral Minaret, Mosque of Ibn Tulun, Cairo, AD 879. Design of spiral tower to left-hand side based upon the Minaret at Samarra, Iraq

Figure 2.1d El Castillo Chichén-Itzá, Yucatán, Mexico, thirteenth century. (Courtesy of Judith Blanc)

The ceremonial stairs leading to the top of El Castillo Chichén-Itzá, Yucatán (Figure 2.1d) signify an awesome and more fearful concept. An ascent leading to human sacrifice on the altar which surmounts the great pyramid. The four flights are for priests and their victims; the gory procedure, which included dismemberment, coated the upper ziggurat with blood, transforming the stonework to 'the colour of a sunset' according to the first Spanish conquerors.[1] This sanguine aspect of religious endeavour is horrendous in terms of both the ziggurat and the slaughter achieved – the main stepped temple which served the capital had a collection of one hundred thousand skulls.

Equally repellent is the gross scale and overpowering symbolism chosen by the Nazis for the vast stadiums in Berlin and Nuremberg, with their rows of serried steps and forbidding rostra. Apologists for Speer will draw parallels with the Colosseum. There can be no comparison in architectural terms between the coarsest pastiche and a masterpiece of Roman monumentality. In practical terms the amphitheatre completed by Domitian is a veritable palace of stairs and vestibules with a remarkable circulation pattern for the 50,000 spectators (Figures 2.1e,f). A seminal stadium for all time.

The vision of the megalomaniac is certainly an element in the creation of stairs and steps and not simply a figment of the apocalyptic imagination of Martin or Piranesi. The obsessive Piranesian complexity achieved in the stairways of the Palais de Justice, Brussels, totally overwhelms the senses and certainly provides the best setting for understanding Kafka (see Figures 5.4a, b, c).

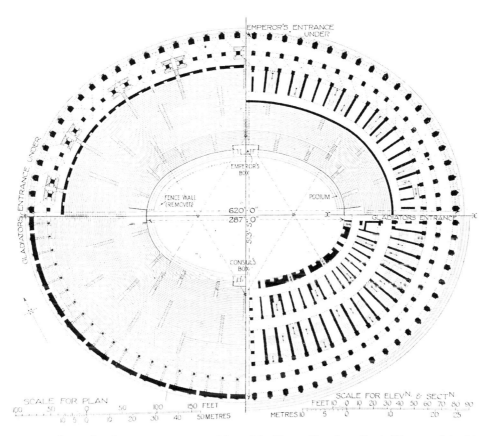

Figure 2.1e Part plan of Colosseum, Rome AD 72 to 82. (From Fletcher, B., A History of Architecture, Batsford, 1945)

Figure 2.1f General view of stairs constructed over brick vaults at the Colosseum

The association of reverence with ascent or with high places has been mentioned briefly and needs expansion. Reverting to Genesis 28:10–28:17, there is the episode of Jacobs Ladder that is purely visionary 'a ladder set up on the earth and the top of it reached to heaven'. Nothing is given about the construction. The notion of high places and divinity is a constant theme in many religions, and the physical climb, step by step, is seen in a holistic manner. The setting of many Buddhist shrines is part of the hill or mountain scenery. A sequential vision is imparted of ramping paths, shrines, continual climbing with the steps becoming harder and harder until the pilgrimage terminates among bare rocks and clouds at

10 Stairs, Steps and Ramps

Figure 2.1g A Jacobs Ladder in landscape terms. Steps called 'the skyladder' at Tai Shan, China (551 BC). (From Schuster, F., Treppen, *Hoffman Verlag, 1949)*

Figure 2.1h Threshold steps in traditional Japanese house

the mountain peak: a Jacob's Ladder in landscape terms (Figure 2.1g). The theme is fully developed in the introduction to External Stairs (Chapter 6).

The other philosophical stance taken from the Orient relates to thresholds – the treatment of buildings at entries. The raised platform, the zig-zag or turned approach, the threshold beam or podium block are all part of the repertoires emanating from vernacular architecture in China and Japan (Figure 2.1h). The infusion of these ideas into western architecture has taken many directions. There are the West Coast designs of Greene and Greene, Neutra, Schindler and Wright. There is the earlier Romantic ideal chosen by Repton who saw the footprint of a building sitting flush within the sward, a 'grace step' of an inch or so separating outside from inside.

Finally there is the inheritance from the Renaissance, Vitruvius was of no help since stairs were not mentioned in the 'Ten Books' except for temple steps to be arranged in uneven numbers. The Greeks could be attributed with the golden rule for tread-to-rise proportions judging by those used in their theatre stairs. These give the customary proportion of twice the rise plus the tread to equal 550 mm to 700 mm. Alberti printed his 'Ten Books' in 1485 though the ideas were circulating in 1450. Stairs are not included in the six elements which constitute a building, however there is advice that stairs should be accommodated within a defined space, ideally a separate room, in the composition. There should be three openings, one at the entry to the stairs and two at the other level. Windows are required for lighting and the space is distinguished by a dome. The solution to such dictates lead to staircase rooms that anticipated the 'Treppenhaus' of the baroque (see Figures 5.2*a*, *b*). The key Renaissance example is the grand staircase that leads to the Biblioteca Laurenziana, Florence designed by Michelangelo in 1523–26 and completed by Giorgio Vasari in 1571 (Figures 2.1*i*, *j*). Here the triple flight fulfils Alberti's promise, assuming an importance of its own in the inner space of the building and thus balancing the library it serves. Truly a foretaste of the more extravagant designs to come from the baroque era and nineteenth-century historicism.

Figure 2.1j General view of Laurenzian staircase, constructed 1571 by Giorgio Vasari. (Courtesy of Kina Italia, Milano, Italy)

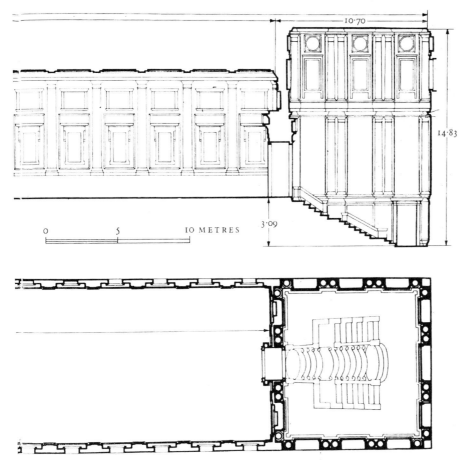

Figure 2.1i Plan and section, Laurenzian Staircase, Florence 1523–26. Designer Michaelangelo. (From Ackerman, J. S., The Architecture of Michaelangelo, *Penguin, 1970)*

2.2 Categories of stairs

The first technical illustration (Figure 2.2.) depicts the various categories of stairs and shows the elaboration that can be achieved by adding turned steps or landings. Floor to floor heights are not always constant for buildings, the strategy in adopting landings and return steps often helps to accommodate differing storeys without increasing the basic stair well. The direct flight or single return version will occupy a greater floor area once the extent of upper landings are taken into account. The popularity of dog-leg or three-turn flights rests with the minimal core dimension required.

The three- and four-turn arrangements facilitate rotating the axis of approach. They also allow permutations to be made in terms of priority, landing by landing and turn by turn.

From the archaeological point of view, traces of stairs or long flights of steps often provide the main clues to the multi-storey building that once existed. Likewise the remains of formal steps or shaped ramps help provide a framework in understanding the relationship between building elements. Space here does not permit a full exploration inspired by the *Comparative Method* of Sir Banister Fletcher,[2] but it is possible to review the salient forms by looking at the various categories of stair in detail, direct, single turn, dog-leg, multi-turn and curvilinear.

2.3 Direct flights

Direct flights of steps are often the most dramatic approach, particularly where they continue the line of movement from one level to the next on the main axis. Egyptian examples often employ stepped ramps due to the complex processional requirements of conveying crowds of people and their artefacts.

A key example is the Temple of Queen Hatsheput, Dêr-el-Bahari (Figure 2.3).

Figure 2.3 Stepped approach to the Temple of Queen Hatsheput, Dêr-el-Bahari, circa 1520 BC

Within some temples internal stairs were arranged as direct flights to enable the priests to ascend to the roof top for services, as at Dendera (Figure 2.4), where spandrel decorations show processions attending the New Year celebrations. These passages rise seven metres and are designed with continuous steps as generous as 50 mm × 500 mm proportion. Adequate lighting is achieved by tapered shafts cut through the walling blocks. The route upwards has a slope cut within the tread thickness so that the effective rise is 100 mm, thus limiting the total length of stepped passage to around 35 metres. The descent is via a four turn stair to save space.[3]

Stairs, Steps and Ramps 13

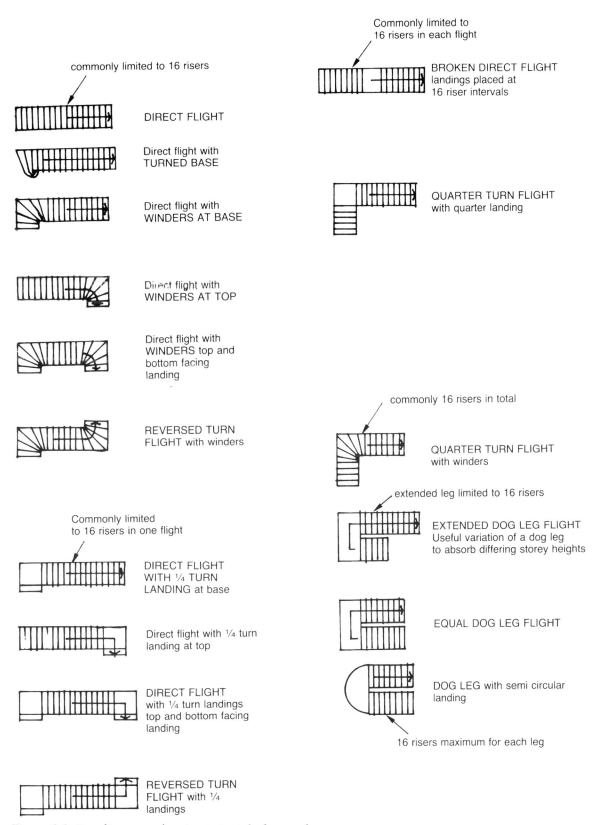

Figure 2.2 Key diagrams showing principle forms of steps

14 Stairs, Steps and Ramps

THREE QUARTER TURN FLIGHT with winders and ¼ turn landing

commonly 16 risers in total

UNEQUAL DOG LEG FLIGHT with dancing steps

EQUAL DOG LEG FLIGHT with dancing steps

commonly 16 risers in total

HALF CYLINDRICAL STRINGS with dancing steps

Emphasis on upper flight — **DIVIDED RETURN FLIGHT** main flight often twice the divided flights

BROKEN DIRECT FLIGHT Landing often used as secondary approach

Emphasis on upper flight (twice lower flight width)

FOUR QUARTER TURN FLIGHT with ¼ landings

RETURN FLIGHT (reduced version of two quarter turns for lesser storeys)

RETURN FLIGHT with two landings

Return flight with most compact form using winders

UNEQUAL RETURN FLIGHT

16 risers max for long leg

DIVIDED FLIGHTS

Main flight often twice width of upper flight

DIVIDED RETURN FLIGHTS (emphasis on lower flight)

extended tread at base

RADIUS STEPS

CYLINDRICAL STAIRS

Cylindrical stairs with landing and **CANTILEVER STEPS** with half circular well

ELLIPTICAL STAIRS (using three centred curves)

landing slab / Landing

OPEN WELL Circular stairs

Landing

SOLID WELL Spiral (steps cantilevered off central column and built between pier and walls)

Figure 2.2 Key diagrams showing principle forms of steps (continued)

Stairs, Steps and Ramps 15

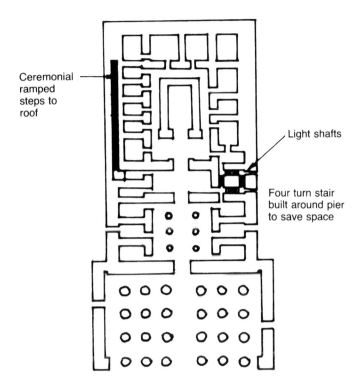

Figure 2.4 Egyptian temple stairs a Key plan of continuous upward flight to the rooftop of the Temple at Dendera, circa 110 BC–AD 68

Figure 2.4b View near roof level, Dendera

Formal steps link the ground floor elements of this temple complex and are designed to demark the ascent to the inner sanctum. The arrangement is common to Egyptian temple architecture, with the rising thresholds devised as part of the immense portals that subdivide the sequence of spaces. The tread to rise proportions provide a flowing transition with the surface 'sunk' into the masonry block strings, giving a flattened triangular form when viewed obliquely. This elegant simplicity is described in greater detail where it is relevant to landscape design in Chapter 6 (Figures 6.12*d* and 6.14*d*).

In Greek temple architecture the division between the public and exclusively religious domain made the external and public approach a crucial aspect in the placing and setting of buildings. A superb example is the Propylaea where ascending spaces lead through the Propylaea and

connecting in turn to the temples on the Acropolis (Figure 2.5).[4] Direct flights exist for the public in contrast to platform steps

Figure 2.5 General view of Acropolis

to demark each temple placement, the stylobate reflecting the scale and significance of each building. The Parthenon has a triple block plinth each 500 mm × 750 mm rising 1500 on the west front. Such a stylobate is difficult to scale and forms the intended barrier to the temple domain. By contrast, Nikè Apterous has a totally human scale with the stylobate profile perfectly proportional for use and related to the temple shrine that it serves (Figures 2.6a, b). Another refinement introduced to the flat site at Pæstum is the banking of the ground for each temple, a subtle device perhaps to hide

Figure 2.6 Steps to the Nikè Apteros, circa 426 BC a General view

Figure 2.6b Detail of Nikè Apteros steps

the spoil from the excavations but which affords a raised, yet natural looking, hill for each element in the composition.

Roman temples, unlike the Greek, were designed for public access with an arrangement that allowed for a congregation as well as altar spaces. The Maison Carrée at Nîmes is typical of the developed plan with a fine direct flight leading to the principal elevation (Figure 2.7). The step profile is repeated in the stylobate and that in turn surmounts the podium walls either side of the entry.

2.4 Multi-turn flights

The purpose in multi-turn design may well have been defensive. The approach stairs in citadels like the Alhambra are strategically planned for right-handed swordsmen fighting in defence. The attackers unless left-handed were under a disadvantage at every landing. The saving in space is another reason for multi-turn flights and a good example is the compact dog-leg and four-turn stairs that serve Pylon towers in Egyptian temples (Figure 2.8).

The most notable internal stair that has survived from antiquity is the 'Grand

*Figure 2.7 Direct flight at main entry to the Maison Carrée Nîmes, 16 BC
a Portico*

Figure 2.8 Compact four-turn stair carved from solid blocks, including spandrel. Great Temple of Ammon, Karnak, 1550–323 BC

Figure 2.7b Detail of steps at Nîmes

Staircase' to the royal apartments in the Palace of Knossus (Figure 2.9). The restoration seen today amounts to a reinforced concrete reconstruction which is partly painted to resemble stucco or timber or else faced in salvaged stone for balustrade and envelope walls. The rebuilding by Sir Arthur Evans[5] has been criticized but at least it is possible to experience the spatial sensation of penetrating five storeys of the Palace by the multiple turns of a finely scaled 'Grand Staircase'. The sequence starts with the roof terraces and then leads the visitor down to the Royal apartments in a specific order of importance. This downward approach is unusual and takes place within a wide hall amply lit by clerestory lights, thus ensuring daylight to the core of the plan.

In the Renaissance such layouts, arranged with an upward approach,

18 Stairs, Steps and Ramps

Figure 2.9 Restored three-turn stairs at the Palace of Knossos, circa 1700 BC

enabled the centrally-placed stairs to be the main instrument in composing plans. A judicious arrangement of landings helped to order plans with primary, secondary and minor circulations (Figure 5.3). Double return versions allowed the axis to be turned left or right, forwards or backwards. A prime example is the Palazzo Municipio in Genoa which has two principal stairs; the first is used to separate the raised ground floor from the street; the second has the crucial role of turning the circulation back towards the important accommodation around the upper gallery (Figure 2.10).

The geometrical progression described takes into account a simple pattern of movement which follows a logical pattern on a clockwise or anti-clockwise route. The Italian Palazzo with its cortile and adjacent stairs leading to the upper galleries is a seminal pattern for country and town houses found along the northern mediterranean. The old town of Ibiza is distinguished by a number of courtyard

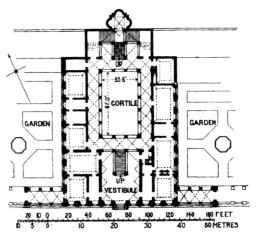

Figure 2.10 Palazzo Municipio, Genoa, 1564. (Lurago) (From Fletcher, B., A History of Architecture, Batsford, 1945)
a Plan

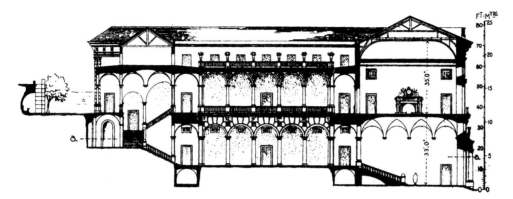

Figure 2.10b Section Palazzo Municipio

houses where the main stairs rise on the wall opposite the entry, as illustrated in Figure 2.11. Here the 'piano nobile' is located to the top right of the picture and overhead, thus maintaining the 'clockwise' access to the ensemble. In more lavish houses, secondary or back stairs exist linking the cellars through every floor to the attic.

It is interesting to find that Palladian villas often relegate the vertical circulation to turret steps hidden within masonry piers. The miniaturizing of the Rotonda (Villa Capra) at Chiswick House generates a meaness in provision whereby the winders are barely a metre wide (Figure 2.12). (A detailed view of the masonry construction involved is set out in Chapter 10). The duplication of stairs within Palladian plans can be explained as solutions to fulfilling the aesthetics of symmetry. However a more practical purpose was the social division of separate stairs for master and servant. The interlocking flights, such as those attributed to da Vinci at Chambord (see Figure 2.18 below), were devised to separate the Royal steps from those of the Court. The notebooks of the artist-inventor reveal a fascination with the geometric

Figure 2.11 Old courtyard house in the town of Ibiza. A seminal example of the Mediterranean courtyard plan

20 Stairs, Steps and Ramps

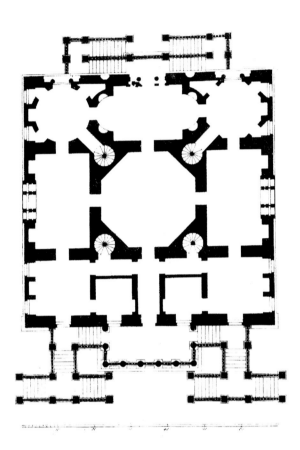

Figure 2.12 Turret stairs at Chiswick House, 1727–29. (Richard Boyle, 3rd Earl of Burlington) (From Summerson, J., Architecture in Britain 1530–1830, *Penguin, 1991, © Yale University Press)*

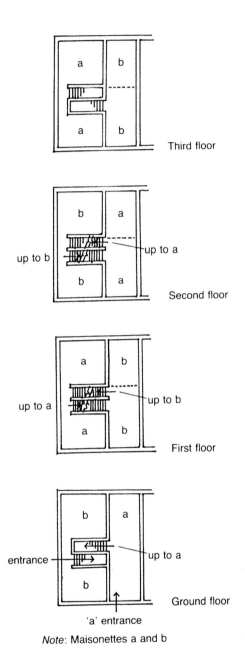

Figure 2.13 Flat plan from sixteenth-century Venice (now Via Garibaldi)

possibilities of interlocking volumes to provide the most compact solution to housing two or more staircases within a single shaft. The double escape stairs shown stacked one above the other in Figure 4.4a have their origins in a da Vinci sketch. Italian elaboration was fully developed in Venetian apartments. In those interwoven plans, the greatest emphasis was upon privacy and security of entrance, with each flat having separate flights from individual front doors at street level. Figure 2.13 demonstrates the complications in the stacking arrangement, but it does make for easier management than contemporary blocks of flats, where twenty or more families may share a common entrance. For comparison refer to the public spaces of plan (*b*) in Figure 3.28.

2.5 Tapered flights

False perspective is a familiar device in landscape design but remarkable examples occur in interiors. Bernini's 'scala regia' at the Vatican (Figure 2.14a). has to be seen as stage scenery to dramatize the passage of the Pope to the Sistine Chapel. The traditional ceremony involves a brief pause as the blessing is given from the top landing to the crowd assembled below. The tapered flight at Venturi's National Gallery extension has no claim to drama other than highlighting a lift lobby. It appears a pointless aberration when designing a principal public entrance for crowds of visitors as opposed to the processional purposed by Bernini. His 'scala regia' is in fact only used for formal occasions, apart from the lowermost connection to St Peters (Figure 2.14b, c). The Scala Santa, as it is also called, was made a punishment for Martin Luther, who was forced to climb the stairs on his knees and to kiss each step *en route.*

Tapering the geometry of spaces or steps changes the sense of distance and enhances the vertical scale of adjacent buildings and sculpture. One of the most successful external designs is the work of Michelangelo at the Capitol in Rome (Figure 2.15a, b, c). The determining factors are the abrupt changes of level between Via del Mare and the Piazza and the further rise to the Palazzo Senatorio, built over the ruins of the Tabularium. The primary route is via a long tapered ramp of steps that terminates between a pair of giant statues by Dioscuri, made more dramatic by the reverse perspective. The arcaded fronts of the Conservatori and the Capitoline Museum are also canted in reverse alignment to enclose the full façade of the Palazzo. The final rise to the 'piano nobile' is made externally with return stairs that link the piazza arcades to the entry portico at first floor. The real magic of the eccentric geometry is captured by the mounded star paving that connects the eye with each element – colonnade, fountain, ramp, stairs and the centrally placed equestrian statue of Marcus Aurelius. Functional requirements such as shelter and vehicles are met by a shaded serpentine path whilst today cars use a winding road tucked away to the side. The surfaces

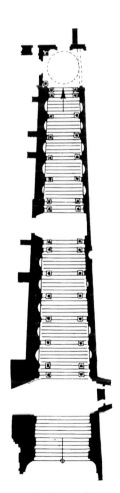

Figure 2.14 'Scala regia' at the Vatican, 1661. (G. L. Bernini) (From Schuster, F., Treppen, *Hoffman Verlag, 1949*)
a Plan

22 Stairs, Steps and Ramps

Figure 2.14b View of 'scala regia'

Stairs, Steps and Ramps 23

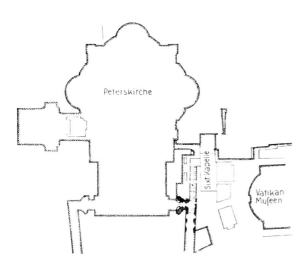

Figure 2.14c Key plan of 'scala regia'

Figure 2.15b View of main flight to the Capitol

Figure 2.15c Piazza paving to the Capitol

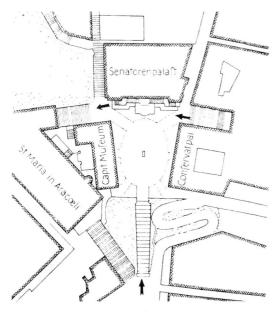

Figure 2.15 The Capitol, Rome, Architect Michelangelo, first phase, 1540–1644, piazza paving. (From Schuster, F., Treppen, *Hoffman Verlag, 1949)*
a Layout plan

are picked out in contrasting materials, dark grey granite with white travertine blocks for risers and for the star pattern to the Piazza.

2.6 Cylindrical and spiral

A basic subdivision exists between winding steps that occupy a circular drum and those that are constructed within a tapering shaft and are therefore of helical form. The latter relate to castle and church towers where the thickness of the turret walls are reduced in the upper storeys. Antonio Gaudi took the Gothic principle to the extreme limits of stability with the four completed towers to the front of the Sagrada Familia (Figure 2.16a) shaped like bottles of hock. The construction involves

24 Stairs, Steps and Ramps

a

b

Figure 2.16 Sagrada Familia, Barcelona (towers to southfront early 1920s), (Antonio Gaudi)
a The completed towers with treads visible between piers
b The upper most portion with cantilever treads and carved ends to form handrail

two rings of piers with the individual treads forming the connecting slabs. Climbing the towers is not for the faint hearted as the outer piers have ventilation slots cut tread to soffit giving unsurpassed views downward to the street below. The inner shaft also forms a perforated wall top to bottom of the helical stairs. The central well is a tapered open void rising through the full height of the tower. The reduction of size is quite awesome with the highest levels having slabs spanning 1 200 m supported by masonry piers as slender as 200 mm. The ultimate portion adopts the cantilever form with the internal well left open while the tread ends are carved to form a welcome handhold (Figure 2.16b)

Cylindrical stairs have been made as free-standing towers placed outside the building envelope. Some of the more ambitious designs permit horse and rider to ride up to the upper floors. This is the explanation for the complex masonry construction at Blois (Figure 2.17) which is built as a twin arcade to support the spiralling steps. It is said to have been influenced by Leonardo da Vinci.

A more ingenious solution inspired by the same artist connects the royal apartments at Chambord. The double circular stairs enable the regal path to be separate

Figure 2.17 External staircase tower, Chateau de Blois, 1515–30

a

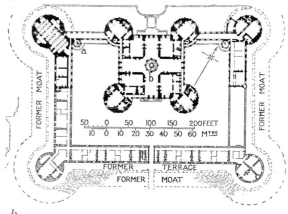

b

Figure 2.18 Double cylindrical stairs at Chambord, 1519–47. (From Fletcher, B., A History of Architecture, *Batsford, 1945)*
a Sketch
b Key plan

from the rest (Figure 2.18*a*, *b*). The magnificent stonework forms the central feature of the cruciform plan and makes a complete contrast to the Palladian ideas of hiding away internal stairs. Another visual advantage at Chambord is the celebration of the cylindrical shape above the roof level where a domed lantern has Italian lines to surpass the giant chimney stacks. The practical purpose is to light the double flight and the core of a plan that is over 42 m in depth. Doubling the flights is arranged with the simple expedient of stacking one above the other on 90 degree turns, the storey heights of 5 700 mm permitting this solution. The da Vinci connection with Chambord is derived from the artist's sketches of four flights nesting one into the other,[6] the actual construction followed in 1526–1533, some years after da Vinci's death in 1519. The masons substituted a double staircase copied from the St Bernard Monastery in Paris.

The helical theme has been used in reverse order for two imaginative interiors. In both cases the space within the stair well has been reduced for the descending

26 Stairs, Steps and Ramps

Figure 2.19 Exit stairs, Vatican Museum

route. First, in historical sequence, the exit ramp from the Vatican Museum constructed in the 1930s (Figure 2.19) and second, the more significant application at the Guggenheim Museum. This surely is an inspired development of the earlier idea and transforms the gallery spiral into the major element of the Art Gallery (Figure 2.20). Purists will complain that paintings need horizontal alignment in the surrounding spaces to avoid conflicting views. Those critical aspects are not paramount when viewing pictures at the Guggenheim, the even lighting and the spacious layout give a balanced quality to the exhibition space, without the sloping lines of the floor becoming obstructive. It is true that the modest scale of the wall area prevents very large exhibits being displayed. A more serious criticism is the low concrete balustrade. It does not prevent vertigo, although placed in a sloping plane and angled away from the ramp. Low sight lines are a Wrightian feature and modestly explained by the master as follows: 'I took the human being at five

Figure 2.20 Gallery spiral, Guggenheim Museum, New York. (Frank Lloyd Wright, 1956)
a View downwards

Stairs, Steps and Ramps 27

Figure 20b View upwards, Guggenheim

feet and one half inches tall, like myself, as the human scale. If I had been taller, the scale might have been different'.[7] He meant the 'scale' of modern architecture!

The curvilinear stair of the rococo is perhaps an equal source of inspiration as gothic spirals. The geometric basis is the setting out of 'dancing steps', meaning an arrangement of diminishing or winding treads which can adapt to non-rectangular plans (Figure 2.21). This illustration is taken from *The Carpenter's Assistant* a typical nineteenth-century manual; similar publications covered the masonry construction that gave the elaborate geometry needed for curving stairs. The rococo basis for the work of Victor Horta can be judged by the elaborations to the main staircase at his home in Brussels, now the Musee Horta (Figure 2.22). The plan shapes are mathematically derived with the middle third of the steps made to follow a regular tread-to-rise relationship. The variation across the breadth of the flight places such designs in a non-

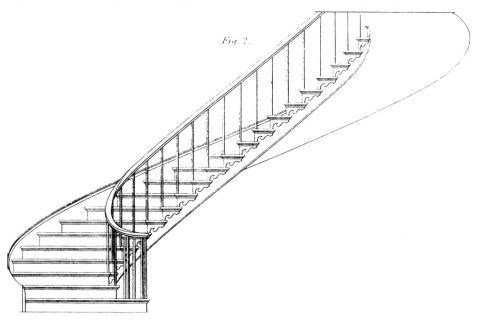

Figure 2.21 Elliptical stair. (From Newlands, J., The Carpenter's Assistant, *Studio Editions, 1990). (See also Figure 7.5f)*

28 Stairs, Steps and Ramps

Figure 2.22 Main staircase, Musee Horta, Brussels 1898. (Victor Horta)

permitted category in terms of public stairs today. There are, however, a number of situations where 'accommodation' stairs are required – connecting flights where alternative 'fire code' stairs exist. Accommodation designs are simply decorative and Horta's imagination fulfils them to full effect. There are few architects today who have managed to create such vibrant, sinuous staircases in steel and timber. The essence of these creations is to capture the movement from floor to floor and to return to the rococo of the eighteenth century (Figure 2.23).

The fusion of construction and spirals has already been mentioned in connection with Antonio Gaudi's work at the Sagrada Familia. His other buildings are all distinguished by remarkable flights of stairs: harking back to the rococo of the late Spanish Renaissance is one of the themes Gaudi brought into the twentieth century. Catalonia seems to have been totally free of petty fogging regulations when Gaudi was creating the Casa Batlló

Figure 2.23 Rococo stairs Schloss Brühl, 1748 (Neumann). (See also Figure 5.2)

(1905–7) and Casa Milà (1905–10). The principal and secondary circulations both follow fully serpentine forms, the concept is roughly elliptical but with a flamboyant double curve provided to the supporting wall. The surfaces to the public areas have majolica or mosaic which reflects the lighting, whether artificial or natural. The secondary shafts are equally complex though the materials are simply stone slabs with painted plaster for the walls, the only extravagance being the wild wrought ironwork to the balustrades.

Describing Gaudi triggers the memory of Casa Milà in 1953 before it had been renovated. The main entrance was not secure in those distant days and it was possible to go through the public spaces and to take the elevator to near roof level to view the famous sculptured chimneys. The escape route led across the surrealist roofscape and then descended on one of Gaudi's forgotten secondary stairs which wound without artificial light down into the bowels of the basement. The form is the pattern already described with double curved flights and every step cut to a differing pattern. It was a matter of feeling one's way step by step. Nobody fell over but there was the lasting impression of dismay in stepping down eight flights, where every foot fall was varied. It was totally exhausting. In conclusion one might learn from Horta who used his wildest dreams for lower storeys, reverting to normality for servant stairs and to those areas where practical considerations were paramount.

2.7 Theatrical stairs

The stepped open air theatres of Ancient Greece provide the source for many modern auditoria like the Olivier at the National Theatre on London's South Bank[8] or the principal space at the new Paris Opera. The most significant reference point is Epidauros (*circa* 350 BC) where Polycleitos designed one of the most perfect Greek theatres, well preserved and now sympathetically restored. The heart of the plan is the orchestral circle that is 20 m across surrounded by 52 rows of seats, bringing the overall diameter to 113 m and holding 15,000 people. The auditory and visual qualities are pre-eminent with a superb relationship between stage and audience. The seat module is derived from the tread-to-riser relationship of the theatre steps, with two risers per stone bench (Figure 2.24*b*). The restoration has revealed an interesting detail – the delineation of seating rows by letters Alpha, Beta, Gama, now recarved on the seat ends.

A key Renaissance example is the Teatro Olimpico, Vicenza by Palladio, where the stage is built in perspective. The setting is permanent and devised as an ideal city with the streets radiating from centre stage. The radial lines are compli-

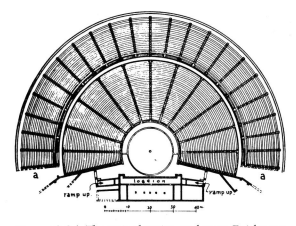

Figure 2.24 Theatrical stairs and seats Epidauros, circa 350 BC. (Polycleitos)
a Key plan

Figure 2.24b View towards stage, Epidauros

mentary to the colonnade and to the curved seating in the auditorium. All these elements tie the geometry of the Greek style benches to the architecture of the proscenium and stage set.

Stairs for the movies are outside the usual building experience but the images hold the imagination. Architects such as Hans Dreir, Paul Nelson and Robert Mallet-Stevens have all participated in Hollywood masterpieces. The best remembered is probably the elliptical flight devised for the plantation home of Scarlet O'Hara in *Gone with the Wind* and designed by William Cameron Menzies. This full scale mansion still stands on the back lot of Universal Studios, though remodelled for countless other epics; other memorable celluloid images today exist only on film. The greatest 'flights of imagination' were the special stages built for the Fred Astaire musicals and which had the most fantastic themes based upon multiple stairs (Figure 2.25).

The theatrical use of stairs is more commonly associated with the vestibules of theatres and vantage points of balconies and landings, designed to show off the audience as they arrive and depart. The most celebrated promenade is that

Figure 2.25 Swing Time, (1936 George Stevens) United Artists, Director Carrol Clarke.

visualised by Charles Garnier – the 'Grand Staircase' at the Paris Opera, already illustrated in the introduction. It is worthwhile studying the plan in detail (Figure 2.26). The area employed almost equals the auditorium and is so arranged that the central well forms a theatre-sized space, overlooked by balconies and foyers on all sides. The stairs rise to quarter landings with the upper flights forming bridges left and right. Wrought iron framing enables the construction to run unsupported floor to floor. The ample proportions and space between solid and void allow the eye to take in the linear movement and enable those climbing to the piano nobile to observe and be observed. In 1981 a successful exhibition was mounted on the Grand Staircase employing the costumes from the operatic

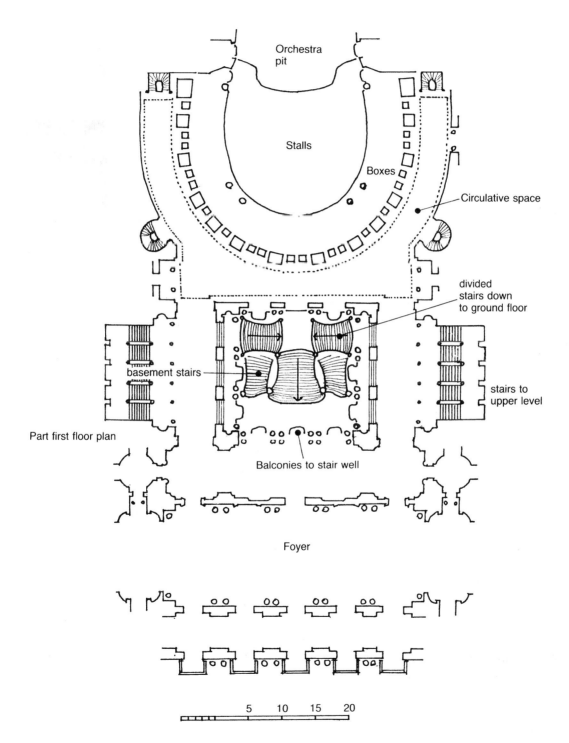

Figure 2.26 The Grand Staircase, Paris Opera House, 1861–74. Key plan at first floor. (Charles Garnier) (See also Figure 1.2)

wardrobes – Tosca leaning over the balcony, Boris Gudinov at the head of the stairs, Cherubino looking for a way out. The impressions (referring back to Chapter 1, Figure 1.2) gave some idea of Garnier's intention to create the most theatrical entrance to the grandest opera house ever built.

It is worth recalling the views of H.S. Goodhart-Rendel when discussing the Paris Opera in the June issue of *Architectural Review* (1949). He tries to recapture the reasoning behind Garnier's design of the Grand Staircase.

Figure 2.27 Foyer to National Theatre, South Bank, 1967–77. (Sir Denys Lasdun and Partners)

A staircase can be regarded either as the rising floor of a building or as a ladder between its storeys. If regarded as a rising floor (as it must be if between walls or around a solid newel) the architectural design of the building must rise with it, keeping pace with its march. If regarded as a ladder it must fly across the architectural design of the building, doing as little damage as possible on the way. Most open well staircases are compromises between the two principles; the architectural design of the 'cage descalier' being complete as though no stairase were there, but adjusted so that in windows, doors, and cornices there are no undesirable collisions between the cage and its content. These adjustments are liable to be awkward, so why not make 'cage' and staircase so independent of each other that none will be necessary? Only where openings in walls have to be reached need the staircase touch the walls at all. And, to emphasize the staircase's independence, why should its flights not curve and twist as livelily as the living load it will carry?

If such reasoning is not unanswerable, the outcome of it in Garnier's hands is beyond cavil. The only criticism of the staircase that has ever occurred to me is that when not burdened with brilliantly dressed people it has a little the air of an empty frame. I do not think this criticism, even if justifiable, is adverse to it. It is intended to be burdened with brilliantly dressed people and looks it. (*Architectural Review*, June 1949, pp. 303–4.)

The staircases in the foyer to the National Theatre, South Bank also form a promenade in space permitting visitors to the twin theatres to enjoy the visual delights of the interconnecting spaces and to sense the connection with the exterior terraces and cityscape beyond (Figure 2.27). The solid concrete balustrades do not however flatter the human form, it could be said in defence of Lasdun's achitecture that today's theatre crowd does not dress for the occasion and perhaps the least seen of their jeans or Bermuda shorts is the best aesthetic decision.

The new 'Grand Staircase' promised for Covent Garden with Jeremy Dixon's plans may bring the focus back to full circle. The intention is a curved space with cylindrical stairs fitted either side of an oval well, the effect will be similar to the interior experienced at the Guggenheim Museum, except that people will be the point of interest rather than walls decorated with paintings. Clearly the final quality will depend upon balustrading and lighting, no mean task when one considers these aspects throughout the five storeys of the intended hallway. (Figures 2.28*a*, *b*, *c*).

34 Stairs, Steps and Ramps

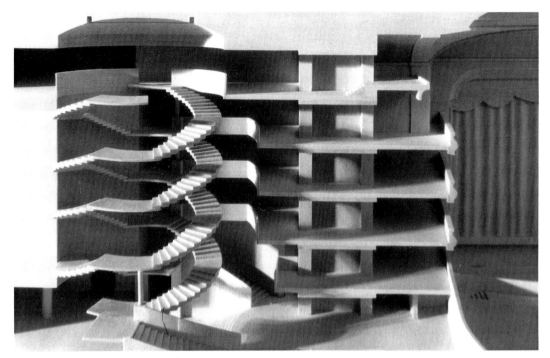

Figure 2.28 Proposed Grand Staircase, Royal Opera House, Covent Garden, 1991 (Jeremy Dixon and BDP)
a Cut away view through model

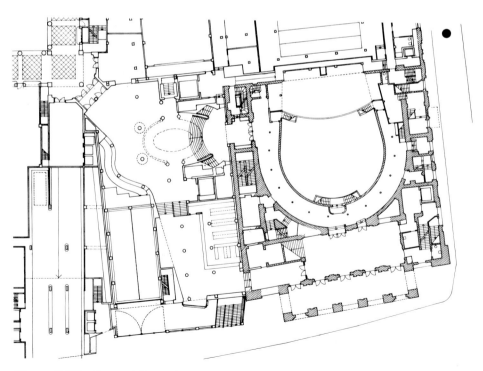

Figure 2.28b Ground floor plan

Stairs, Steps and Ramps 35

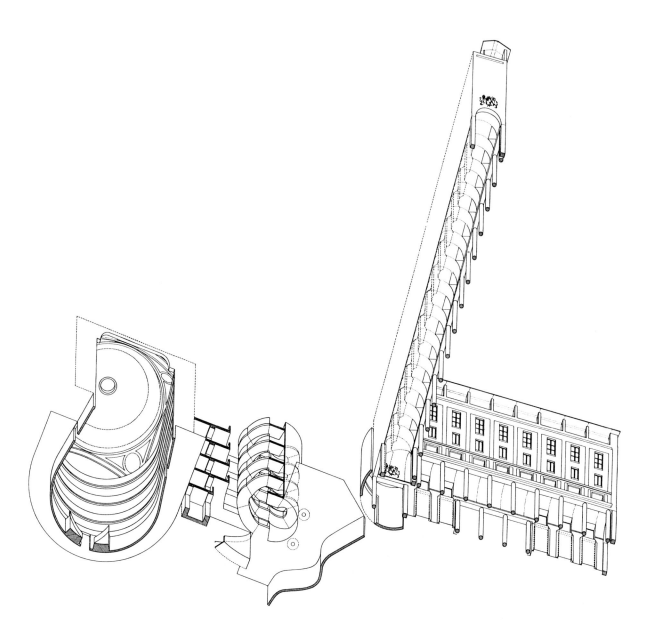

Figure 2.28c Axonometric (looking upwards)

References

[1] *The Rituals of Dining* by Margaret Visser (1992) Viking. A recent book which sets forth the butchery and cannibalism that pervaded the Aztec religion and has eye witness accounts from the sixteenth century of ziggurats in use as abattoirs.

[2] *A History of Architecture on the Comparative Method*, by Sir Banister Fletcher. A remarkable collection of illustrations and text that ran through many editions starting in 1896. Re-written with fresh drawings and published by Butterworth in 1987.

[3] There is evidence from the wallpaintings that stairs were used separately to serve differing roof temples. Today with tourist crowds it has been found safer to use the unbroken straight flight for the ascending route.

[4] *Towards a New Architecture* by Le Corbusier, chapter entitled 'Architecture' (III Pure Creation of the Mind). Authorities today claim that the steps leading through the Propylaea are probably Roman in origin, the Greek construction being a ramped pathway.

[5] *The Find of a Lifetime* by Sylvia Horwitz. An account of Knossus, its discovery and rebuilding. Evan's first repairs were carried out in timber and stone just as the original construction. Decay quickly destroyed the pine trunks, hence simulation in painted reinforced concrete.

[6] *Apprenticed to Genius* by Edgar Tafel, p. 50.

[7] These sketches of quadruple stairs by Leonardo da Vinci are kept in the library of the Institut de France MS B fol. 47.

[8] The parallels between Epidauros and the Olivier Theatre are drawn upon by Sir Denys Lasdun in his book *A Language and a Theme* and cross referenced in Sir Peter Hall's *Diaries* (p. 165).

3 Domestic stairs

3.1 Generic plans

The layout of both houses and flats can be summarized on a generic basis,[1] i.e. the allocation of minor and major zones for servant spaces and habitable areas respectively. Broadly speaking, the minor zones are associated with stairs, bathrooms, kitchens, utility areas as well as with small single bedrooms or studies, while major zones are made up of the living rooms and the larger bedrooms. If the assumption is made that all rooms shall be accessible from the common stair hall – to avoid using habitable areas as passageways – plan variations are effectively limited for the smaller house, to two basic forms. First, the 'Universal' plan (Figures 3.1a, b) and its modification for long fronted or 'L' formation (Figures 3.1c, d). Second, for wider frontages, is the 'Double Fronted' plan and its wealth of central core layouts (Figures 3.2a to g).

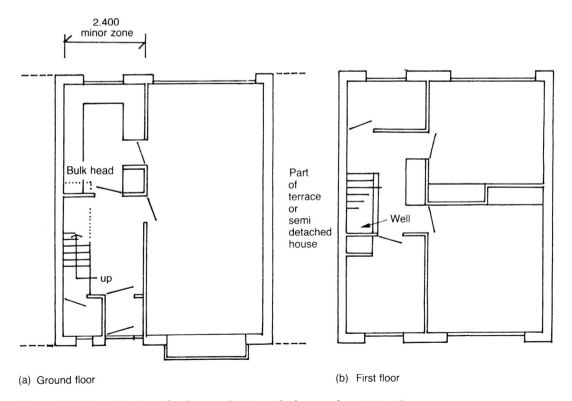

Figure 3.1 Generic plans for houses (universal plan and variations) a Universal plan, ground floor and b first floor

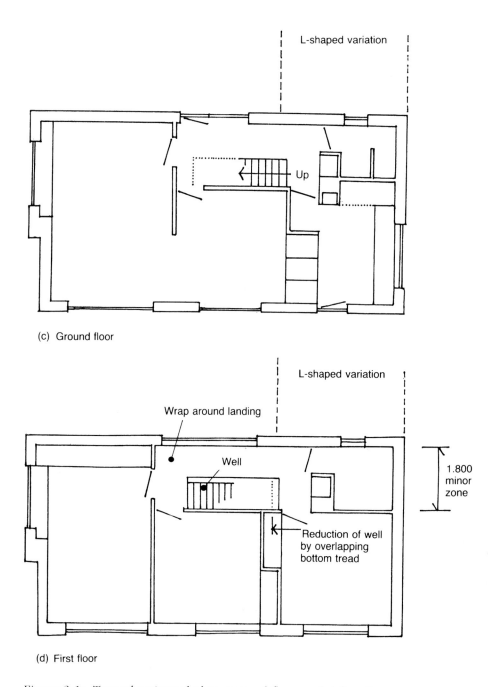

Figure 3.1c Turned universal plan, ground floor and d first floor

The actual dimensions allocated for the minor and major elements depend upon the modular dimensions adopted. The common arrangement for stairs use twin bays in one direction. It is then customary to provide twin modules of 900 mm, giving 1 800 mm for the width of cores. 1 200 mm is a common module in system building and which provides core widths of 2 400 mm (Figure 3.1e).

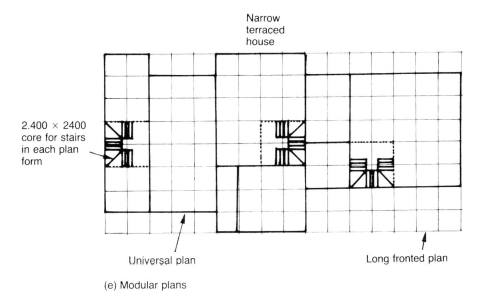

Figure 3.1e Modular plans (1200 grid)

In Britain the 1970's attempt at metrification led stair manufacturers to set preferred heights for floor-to-floor increments in housing, and as a result standardized flights are now made for 2 600 mm storey dimensions. The same process exists in continental Europe and in North America, though to differing standards. There is much to be said for Aalto's modular process in making staircases which related to one centimetre increments without the crippling interference of governmental worthies.[2]

3.2 Staircase options

3.2.1 Direct flights

The single direct flight provides the cheapest constructional solution. If laid out across the plan it will give the most compact answer (Figure 3.2a). This minimal aspect explains its use in cottage architecture where there are no halls or lobbies. In the simplest plans, the staircase would rise directly opposite the street door with the accommodation arranged

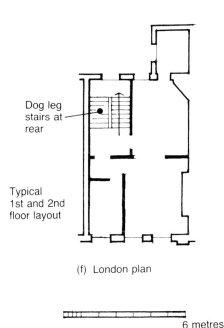

Figure 3.1f London plan (based upon 7 Frith Street, Soho. (Cited by Ware in 1756 as the plan form of the 'common house')

40 Stairs, Steps and Ramps

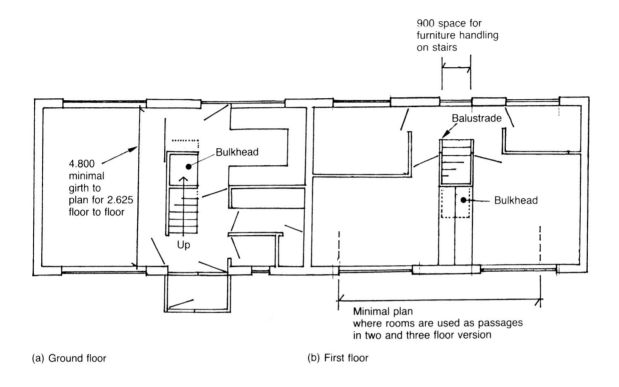

Figure 3.2a Generic plans for double fronted houses and b Cottage plan

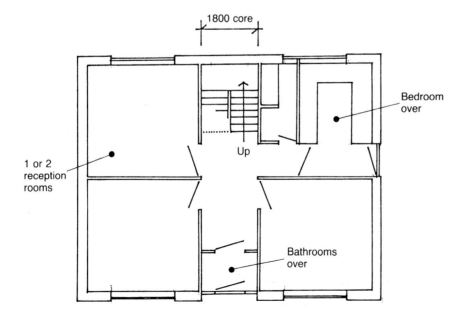

Figure 3.2c Double-fronted Georgian plan

Stairs, Steps and Ramps 41

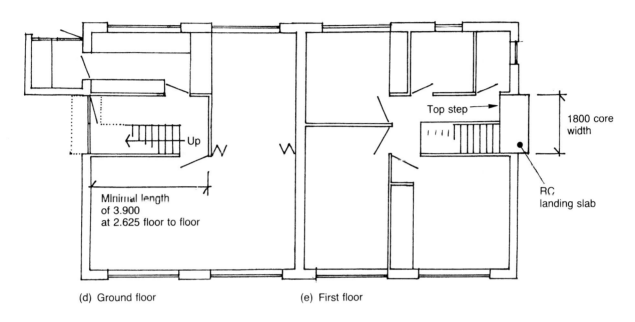

Figure 3.2d and e Turned double fronted plan

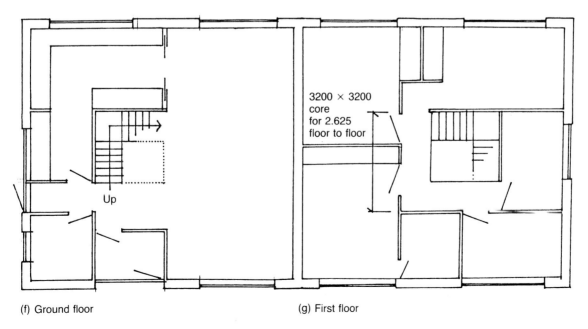

Figure 3.2f and g Central core plan

to one or both sides. In these cases, living areas or bedrooms were used as 'through rooms' to keep the envelope walls to the absolute minimum, often resulting in overall dimensions as small as 30 m^2 per floor (Figure 3.2*b*).

The same direct flight set longitudinally produces a 'wrap around' hall at each floor level, a slight saving on this extravagance can be achieved when the well is returned over the second riser – providing the required headroom is available (Figure 3.1*d*). A solution by Walter Segal (Figures 3.2*d, e*) gives an oriel landing, which shortens the hall by 450 mm and also provides a sheltering porch at the entry. The single top step does not comply with current National Building Regulations in the UK. In times past, non-conforming use with the meanest winders, coupled with 200 × 200 treads and risers, enabled Dutch designers to reach the ultimate levels of parsimony. The stairs in old Amsterdam houses have ladder-like steps more appropriate to ships than residences.

Winders to the mandatory taper are permitted and will reduce hallways by 900 mm, but it is better to keep these at the base of the stair for safety. It is worth remembering that a quarter turn of winders equals four treads and that the arrangement can successfully reorder a hallway to give more space for furniture (Figure 3.3). For variations with winders refer back to the details in Chapter 2 (and see Figure 2.2). Purists will say that it should be possible to design a flight without tapered or winding treads; pleas are put forward in relation to comfort and safety in use, particularly for the physically handicapped. The straight flight and a full length handrail are much preferred by them to coping with awkward treads, which are too far from newel posts or from an effective handrail (refer to Figure 3.4). Having said that, it is salutary to recall Mrs Schroder (of Rietveld–Schroder House fame) who daily climbed up and down a ship-like set of Rietveld's tortuous winders for over 65 years![3] (Figure 3.5).

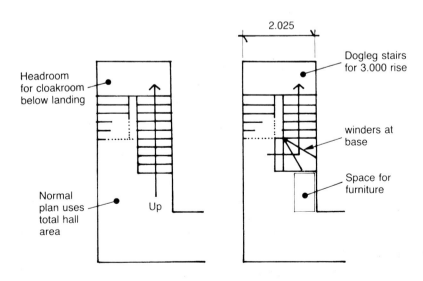

Figure 3.3 Turning stairs to improve a hallway, Redinton Road, 1870s. (Philip Webb)

Figure 3.4 Standard Danish stair installed in the 'Marchesi' System Built House, 1981

Figure 3.5 Tortuous winders in Reitveld–Schroder House, Utrecht, Holland, 1924

Figure 3.6 Benn Levy House, Chelsea, 1935–6. (Gropius and Fry)

There is little doubt that the straightforward layout favoured by Gropius and Fry at the Benn Levy House, Chelsea (Figure 3.6) provides an inspirational example from the 1930s. The fine detailing in steel, timber and glass by Jack Howe has stood the test of time and illustrates the way generosity in plan gives rise to more comfortable proportions. The return landing at the base assists the spatial concept and shifts the main circulation from a collision with the open string. The open composition increases the spaciousness of the interior unlike the traditional approach of infilling below the stairs with cupboards or a lavatory, which brings the space back to corridor size. Another advantage in the detailing at Chelsea is the turned treads at the foot of the stair which divert the line towards the hallway.

3.2.2 Dog-legs

The dog-leg pattern is ideal for double fronted variations (Figure 3.2c). It is also the basis of the 'London' plan that dates back to the 1660s[4]. This 'Universal' layout (Figure 3.1f) places dog-leg flights at the rear of the house so that all rooms are accessed from the common stairs. Variations in storey heights are accommodated by lengthening the stair hall or by the use of winders. The ground floor usually has extended steps towards the entry passage, with the basement route neatly tucked in below. The visual aspect is comely, the best historic examples are probably the seventeenth-century pieces, constructed from oak, and still in existence in the Inns of Court. Further consideration of stone and iron versions are given in Chapter 10.

Replacing the half landings with winders can reduce the floor opening to 1 800 mm × 1 600 mm (Figure 3.4). The area saved is added to the circulation space and ensures a wider range of options for locating doorways at all floor levels. It can also help the standard Universal Plan by reducing the stair hall to a more cubic shape that can be placed centrally on the party wall (Figure 3.1e). Furniture handling on dog-leg stairs is not a problem providing that the newel posts have been terminated close to the balustrade line. Similar stairs built within storey height balustrades create insoluble problems for furniture removal. Another problem area are 'scissors' type stairs constructed within masonary shells, a popular method from the 1960s and 1970s for maisonettes in public housing. The central spine wall needs to be scalloped either end to give a turning circle for handling bulky furniture (Figure 3.7). The tight Amsterdam plans with steep double winders already referred to are the reason why old Dutch houses sprouted gable pulley blocks so that beds, chests and wardrobes could be hauled up externally and passed through upper windows. Scotland's 'turret' houses had the restriction of helical or multi-turn flights with a central pier. Here the problem was solved by making the furniture on site, room by room and leaving it there for posterity!

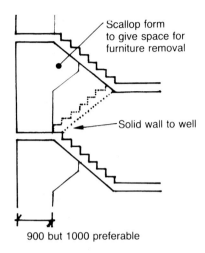

Figure 3.7 Scissors stair with scalloped spine walls

3.2.3 Domestic spiral and cylindrical stairs

The decorative and sculptural quality of spiral stairs is no doubt the reason that so many kits or standard spirals in concrete, steel or timber are now available. A contributing factor is the ease with which computer aids can speed up the designing of such features to fulfil specific site requirements and to meet the onerous restrictions of the current Building Codes. An outline is given in Chapter 11, it will suffice at this stage to confirm that 1 800 mm diameter is the minimum drum for spirals in private houses. This dimension assumes a clear headroom of 2 000 mm. The diameter can be reduced to 1 400 mm for a stair intended for occasional use with access to one room or balcony, etc. The normal diameters of 1 800 show no saving in well dimensions below minimal dog-legs and winders. There is however the visual consideration where the spiral has a greater advantage (Figure 3.8), and the fact that the design cannot be compromised by boxing in for

Stairs, Steps and Ramps 45

Figure 3.8 Domestic spiral stair, Highbury Terrace Mews, London, 1971. (Peter Collymore) (Courtesy of Bill Toomey)

understairs cupboards or with lavatories for pygmies. The design problem has been greatly eased by standardized components and particularly those with threshold arrangements at each access point. Makers offer a range of riser proportions so that quarter landings can be neatly stacked one over the other to cover a choice of storey heights. Differing heights can be accomplished by adjusting landing locations (Figure 3.9). The easiest answer is to employ an open quadrant facing the main circulation, that can be opened or closed according to the number of treads needed as this will produce a more consistent solution. Other methods involve turning the landing segment quarter by quarter where differences in storey height occur. The quarter landing shape makes for an easy transition from floor space to spiral geometry. Gaudi gave short shrift for safety in his designs, access on and off came from a single step befitting the agility of mountain goats (Figure 3.10).

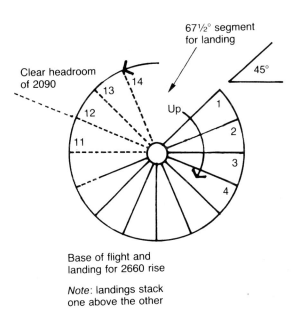

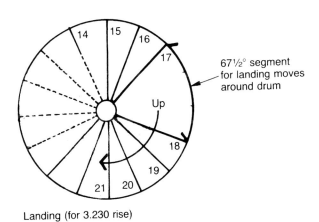

Figure 3.9 Landing locations

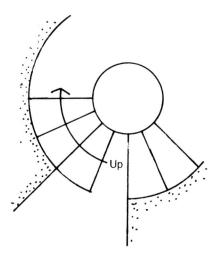

Figure 3.10 Detail of Gaudi stairs with minimal landings

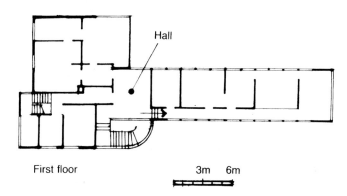

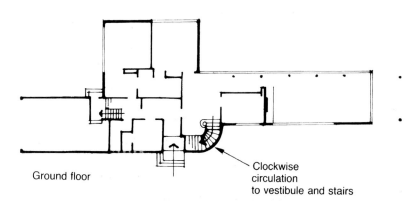

Figure 3.11 'Shrublands', Chalfont St Giles, 1935. (Mendelsohn and Chermayeff)

Cylindrical stairs give a more spacious effect whether placed free standing or within the shell of an enclosing drum. Contrasting the curved and straight line is a feature of early modernism. Mendelsohn and Chermayeff created a seminal house of the thirties at 'Shrublands', Chalfont St Giles, where the swept line of the stair is placed at the crucial break in the plan (Figure 3.11). The elegance of the solution is extended into the detail of the swept platform at the base and to the helical handrail that terminates the balustrade. Another example of this genre is the work of Raymond McGrath at St Annes Hill, Chertsey (Figure 3.12), where a segmental vestibule embraces a cantilever cylindrical stair to the left-hand side. Such exercises in imagination raise expectations for the scenario at first floor. The Mendelsohn theme creates a fine hall with two-way views as well as access to the master suite and guest wing in descending order. McGrath's dramatic layout leads to the circular master bedroom that is the hub of the eighteenth-century landscape garden beyond.[5] Both of these large houses were twin stair

Figure 3.11c View of ground floor flight, Shrublands, Chalfont St Giles

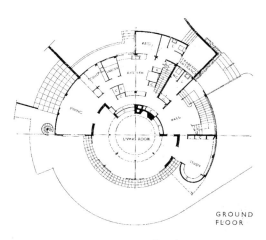

Figure 3.12 St Annes Hill, Chertsey, 1937. (Raymond McGrath)
a Ground floor plan

designs – one for the masters and one for the servants – the secondary stairs were mundane affairs tucked away in the corner of the plan. The social mores of the time dictated these concepts, though the 'gay' lifestyle at St Annes Hill created special conditions whereby the master bedroom had a circulation totally separate from the servants. The origins of white concrete cylindrical stairs must surely be the creations of Le Corbusier and Pierre Jeanneret in their heroic period of the 1920s. The free standing setting within the Villa Savoye raises the stair into the vantage point, with slots cut in the drum wall to reveal the building interior (Figure 3.13*a*). The famous ramp is the ultimate pivotal zone which allows the whole space to unfold from ground floor to roof terrace (Figures 3.13*b*, *c*). The structural role of stair drum or ramp framing is used to the full. Ernö

48 Stairs, Steps and Ramps

Figure 3.12b Hall and stairs St Annes Hall, Chertsey. (Courtesy of Norman McGrath)

Figure 3.13 Villa Savoye 1929-31. (Le Corbusier and Pierre Jeanneret). (Reprinted by courtesy of Charles Jencks).
a Staircase drum   b Stair and ramp

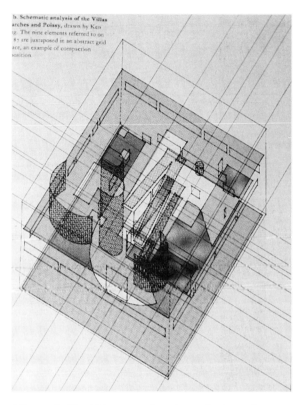

Figure 3.13c Diagram showing relation of stair and ramp

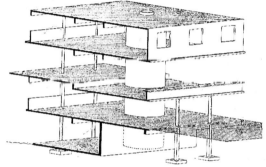

Figure 3.14 Use of stair walls in structure, Willow Road, London NW3, 1937. (Ernö Goldfinger)

Goldfinger borrows the same theme for the half circular flight at Willow Road. Here the reinforced concrete floorslabs run cross wall to cross wall with an intermediate support on the reinforced concrete stair drum that is placed centrally to the depth of the plan (Figure 3.14 and 9.1e).

3.2.4 Modernism and domestic stairs

The key role of the stair within the 'open' house plan has already been alluded to in the work of Le Corbusier at Villa La Roche-Jeanneret (Chapter 1) and with Villa Savoye (Figure 3.13). This concept of the open stair within the open plan is the hallmark of the Corbusian ideal. The initial designs for the Dom-Ino skeletal house (*circa* 1914–15) demonstrated a framed structure with reinforced concrete dog-leg stairs spanning between cantilevered landing slabs (Figure 3.15a).[6] The development of this concept through the designs for Maison 'Citrohan' arranged the stairs as direct flights to save frontage, with the final version built at the Weissenof exhibition, Stuttgart in 1927 (Figures 3.15b, c). The significance of these layouts is the way that dining and living areas are deployed as circulation zones without the traditional separation into rooms and stairhalls. Compare the conventional plans in Figures 3.1a, 3.2a, b, c.

The modernist device is to use the stair as a freely placed feature to liberate

50 Stairs, Steps and Ramps

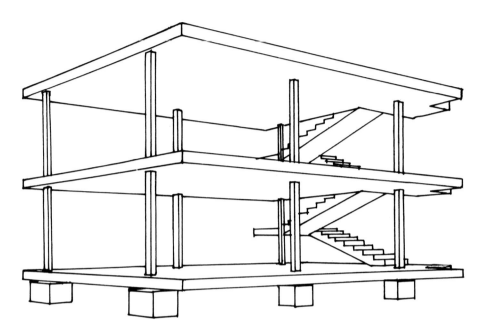

Figure 3.15 Dom-Ino Skeletal House (circa 1914–15). (Redrawn from an early sketch by Le Corbusier)

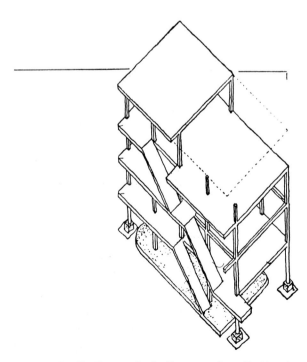

Figure 3.15b Frame for built example at Stuttgart, 1927. (Le Corbusier and Pierre Jeanneret)

the spacial development of the interior. An example of this doctrine is the re-modelling of a pair of houses in St Leonards Terrace, Chelsea by Richard Rogers and Partners. This design opens the traditional vertical box-like London home into a series of large horizontal spaces.

The major element is a double height apartment on the first floor with a projecting mezzanine forming a sleeping balcony. Access within the apartment is made via a staircase running on the diagonal (Figure 3.16a). The use of the longer dimension for ascent increases the sense of space and enhances the fine proportions. The backdrop is a well fenestrated façade comprising twelve windows looking towards Burton's Court. The plan of the whole conversion (Figure 3.16b) has a well contrived sub-division between staff facilities at base-

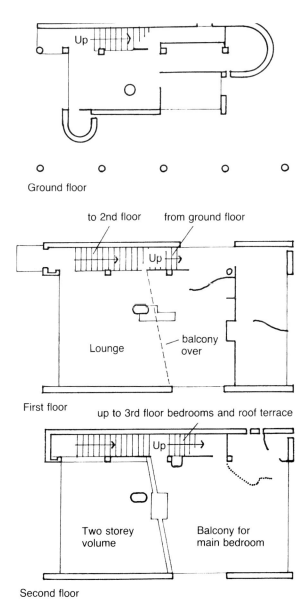

Figure 3.15c Plan arrangements for house at Stuttgart, 1927. (Le Corbusier and Pierre Jeanneret)

ment level, the grandparents' flat at ground floor level and the children's territory zone at the upper levels, with the parental zone of the grand apartment sandwiched between. The principal vertical circulation, top to bottom, is a springy steel spiral staircase which rises in a self-contained well to bypass the principal living area. The subdivision allows for multiple use in the reverse order to the Goldfinger home, which took the idea further and eventually accommodated four generations of the family floor by floor. The youngest family in the garden basement flat, Ernö and Ursula at ground and first floor whilst grandma Goldfinger ruled the second floor.[7]

The Hopkins house in Downshire Hill also shares the spacious quality of open planning, with the stair sited within the middle zone of a nine compartment plan. The planning grid is based upon a minimal steel framing module of 2 × 4 m, the core, that includes bathrooms and the spiral stair, is 2 m wide while the principal rooms towards the garden and street occupy a double bay of 4 m each (Figure 3.17a). The steel spiral forms a skeletal structure in a minimal tube within the main circulation space (Figure 3.17b). Flexibility in the layout allows partitions to be added as family requirements change, with the garden level split into flatlets when the children are older.

The thematic response can best be seen in the cantilever spiral that is at the heart of the Jencks' house. Here each tread fits one to the other like the spiralling of a conch shell. The depth and height are signified by a mosaic telling of the underworld set in the basement floor and by the cascading light from the domical 'eye' to the sky. Each step is numerated to tell of the ascent, the whole movement of the flight being captured by finely made sinuous handrailing. The oval plan is masterly with spreadeagle treads that lead off into the landings. A Gaudiesque invention, this time with no danger to limb (Figure 3.18).

Figure 3.16 Rogers apartment, St Leonards Terrace, London, 1985. (Richard Rogers and Partners)
a General view of 2 storey volume with stair on diagonal

3.2.4 Ladders and steps

Ladders and steps are special forms of a steeper design; they are permitted for particular purposes, e.g. occasional access to roof spaces or service rooms. Loft ladders are generally collapsible or operate with a trapdoor mechanism. The space limitations are the transverse run of the equipment within the roof for the extending pattern and the area needed when the ladder is erected, both the foot space and the girth needed to walk around the obstruction. A rough guide is given in Figure 3.19. Folding scissors ladders exist but these are not as stable as telescopic.

Fixed steps have the advantage that access is permanent and guard rails are fitted at the upper level; in addition non-mechanical items can fit site dimensions more easily. The summary of the British Regulations for long ladders are given in Figure 3.19. Half steps are also acceptable and are more comfortable to the tread than narrow rungs. Ferdinand Kramer (Figure 3.20) patented his version in 1928 but such ideas are not very different from the carved log ladders developed by the Vikings. In principle, ladders and Kramer's patent raise angles of ascent to between 60 and 70 degrees. The construction of the latter can be made in pre-cast concrete, steel or timber; standard flights can be obtained from Denmark and Italy (Figure 3.21)

Stairs, Steps and Ramps 53

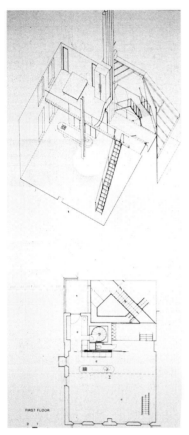

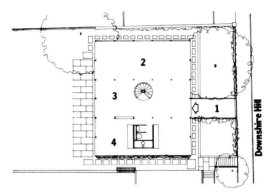

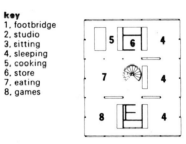

Figure 3.16b Rogers apartment, layout plan (showing two stairs)

Figure 3.17 Hopkins House, Downshire Hill, London, 1976. (Michael and Patty Hopkins) a Layout plan

Figure 3.17b Hopkins House, general view of spiral stairs at core of plan

54 Stairs, Steps and Ramps

Figure 3.18 Stairs at the Jencks House, Labrooke, London, 1986. (Joint designer Terry Farrell and Co.)

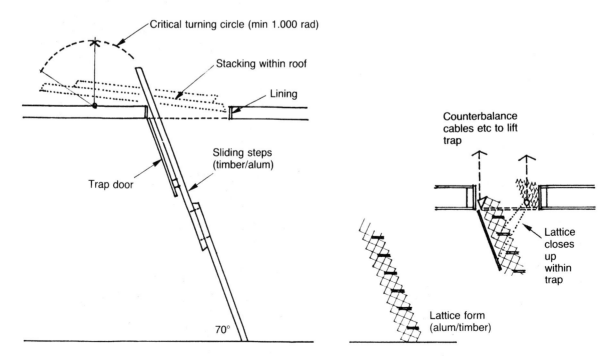

Figure 3.19 Loft ladder principles

3.2.5 Other forms

Three-turn forms take up more space than dog-leg or spirals, however the openness of the central well can enhance the natural lighting given by the upper landing windows or roof lights. One source of these ideas is the vertical core that serves Ottoman domestic architecture. The structures are generally timber post and beam with brick infilling. The treads are often corbelled stone braced by continuous newel framing. Wreathed handrails are replaced by panels of slats with individual handrails fitted between the verticals (Figure 3.22a). A modern derivative by Walter Segal, given in Figure 3.22b, often serves as the internal circulation for maisonettes.

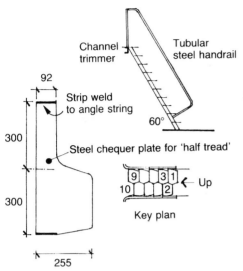

Figure 3.20 Details of early patent (1922) by Dr Ferdinand Kramer

Figure 3.21 Proprietary ladder steps. (Courtesy of Space Saver Stairs (UK) Ltd)

Figure 3.22 Three-turn stairs a Ottoman Merchants House, Cairo, circa *1500*

Figure 3.22b Equivalent to Ottoman design developed by Walter Segal (Flats in Compayne Gardens, London, 1965)

Figure 3.23 Vertical or thermometer window featured in the rear elevation of the Winslow House, Chicago, 1893. (Frank Lloyd Wright)

3.2.6 Elevational considerations

Central siting for stairs, or locations by internal walls, provide no elevational problems apart from the provision of adequate windows to light the steps. This aspect used to be covered in the National Building Regulations with regards to both artificial and natural lighting. Flights placed against the external wall look dramatic behind extensive glazing but make for difficulties with internal window cleaning (see Figure 3.11 above to visualize the problems). Windows that step up with the flight may be the answer but often appear awkward externally. The vertical or thermometer window is another dubious invention (Figure 3.23).

3.3 Stairs and lift cores in flats

3.3.1 Historic background

Dwellings stacked one above the other with staircases placed externally or internally reveal layout patterns little altered by history. The historic and social role of flats is dealt with here in some detail to help explain the crucial aspect of common stairs in their design.

Multi-storey flats date back to Roman times. The remains at Ostia reveal tenement structures that once rose five storeys, with access via internal passages and three-turn stairs, and built to a pattern similar to that adopted by the Peabody Trust in London in the 1870s. Viewing the Ostia ruins confirms the author's view that housing the urban proletariat has made remarkably little progress in the past 2000 years (Figure 3.24). Other examples from medieval times still exist in the cities of the Yemen. They are built as free standing towers of mud brick rising ten or more storeys.

Figure 3.24 Flats in Ostia (reconstruction)

Open balcony access with external turret stairs appears in medieval France and Venice. The pattern is illustrated in the well-known plans by Serlio in his seventh book. A derivation is used in Heriot's Hospital in Edinburgh (Figure 3.25a), with corner turrets and balconies to the courtyard. The form is also utilized in the development of the old town in seventeenth-century Edinburgh (Figure 3.25b). The reasons for external approach probably relate to basic hygiene since the balconies or open stairs led to privies in the outside walls as compared to more advanced Roman methods (*circa* AD 100) where a lucky few enjoyed bathrooms with water pipes fed from roof level aqueducts.[8] 'Walk up' flats were eventually supplanted in the nineteenth century by multi-storey blocks served by hydraulic lifts within internal cores. Such construction was built exclusively for the middle and upper classes.

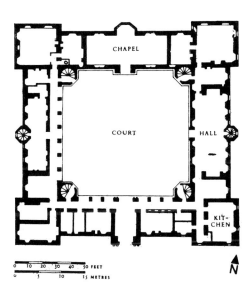

Figure 3.25 Balcony access and stairs in seventeenth-century Edinburgh a Plan of Heriots Hospital, 1628. (From Summerson, J. Architecture in Britain 1530–1830, *Penguin, 1991, © Yale University Press)*

Figure 3.25b Typical tenements behind Royal Mile, Edinburgh (seventeenth century)

London's first street of continental style apartments was Victoria Street (laid out in 1852–71).[9] One of the largest developments was constructed in the 1870s close by St James Park station as Queen Anne's Mansions. Typical plans of a thirteen storey block are given in Figure 3.26, showing a lift and main stairs for tenants with back stairs for servants and tradesmen.

The pattern of balcony access with semi-open stairs dominated low-cost flatted housing in the UK and the Continent throughout the later nineteenth century right through to post-war construction after 1945. This was partly due to the economy of construction with common stairs, spaced as far apart as 48 m, serving perhaps eight flats at each landing.[10] Fire fighting was also easier to organize via external balconies and from staircases situated in the open air. Enthusiasts for balcony access would talk of constructing the social idea of 'streets in the sky'. The reality is often the most unpopular type of dwelling where privacy, quietness and safety is hard to come by. Idealists in the nineteenth century looked upon the new urbanism of balcony access flats in an entirely different way. They saw such layouts as controllable, secure and far superior to the city slums that they replaced. A key figure was Charles Fourier who, in the 1850s, proposed agrarian communes, which were termed a 'phalange', each master block housing 1600, the same number chosen by Le Corbusier for the Unite d'Habitation at Marseilles.[11] A key example of Fourier's dream is the atrium of Familistere at Guise, France (1859–70). The four blocks of balcony access flats look towards a glazed central space, securely guarded by concierge and superintendents. The pattern was copied elsewhere in Central Europe (Figure 3.27). The stairs are usually placed at corners between the blocks with access controlled via the lodge of the concierge at street level.

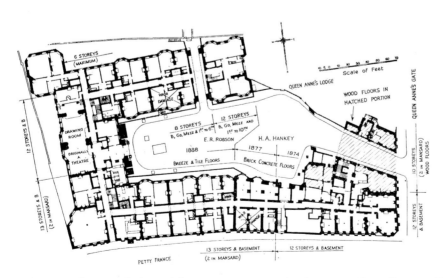

Figure 3.26 Typical plan of flats at Queen Anne's Mansions, Petty France, London with central stairs and hydraulic lift, 1870s. (E. R. Robson – also famous for London School Board Work). (From Hamilton, S. B., Bagenal, H. and White, R. B., A Qualitative study of some buildings in the London area, HMSO, 1964)

Figure 3.27 Nineteenth-century flats in Budapest on the Familistere pattern, with balcony access and corner stairs

The post-Le Corbusier version[11] is a central passage within the core of the building which, under non-existent management, is the worst place to come home to. The description by Dr David Widgery in his book *Some Lives* gives the true horror found in many of London's antisocial flats and maisonettes. The key passage is:

> these antisocial features are even more prominent in the low-rise, high-density estates which succeeded the tower blocks. Typically their pattern is of interdigitating split-level maisonettes, approached by internal corridors without natural light. These corridors become infernal with regularly smashed light fittings, scorched, gouged and sometimes stripped-out lino-

leum, heavily embossed wall surfaces and stale smells. This handiwork is usally executed by passers-through who utilise the corridors and stairwells as meeting spots, skateboard and mountain bike runs, or glue- and heroin-sniffing stations.

While the conventional street gives the residents the psychological advantage of overlooking passers-by whose route and purpose are defined, the internal corridors give the initiative to the outsider. People soon get fed up of opening their (internal) front door in order to identify the noise-maker and instead live in a state of siege trying to ignore the public procession through the middle of their homes.

Direct access to all flats from a central core served by main stairs, lift and secondary fire stairs (if needed) are easier to manage (Figures 3.28*a, b, c, d, f*). Such arrangements have the advantage that noise can be isolated from the quieter zones of the surrounding dwellings, a point of detail excellently illustrated in the Victoria Street flats of the 1870s (Figure 3.26 above).

3.3.2 Generic forms

A generic analysis is given in Figure 3.28, the basic subdivision between single and double staircase forms depending upon Fire Brigade requirements. The fire service must have direct access to a single staircase via a window or balcony point. The lobby is the common method of separate downward escape from the risk of smoke (burning flats or refuse ducts). Tall flats (above Brigade ladder height) are often designed with continuous balcony fronts and external steps to ensure that escape from each dwelling can be made safely via an open air route to an assembly point, which is reachable by Brigade equipment. More detailed consideration of means of escape in the case of flats as

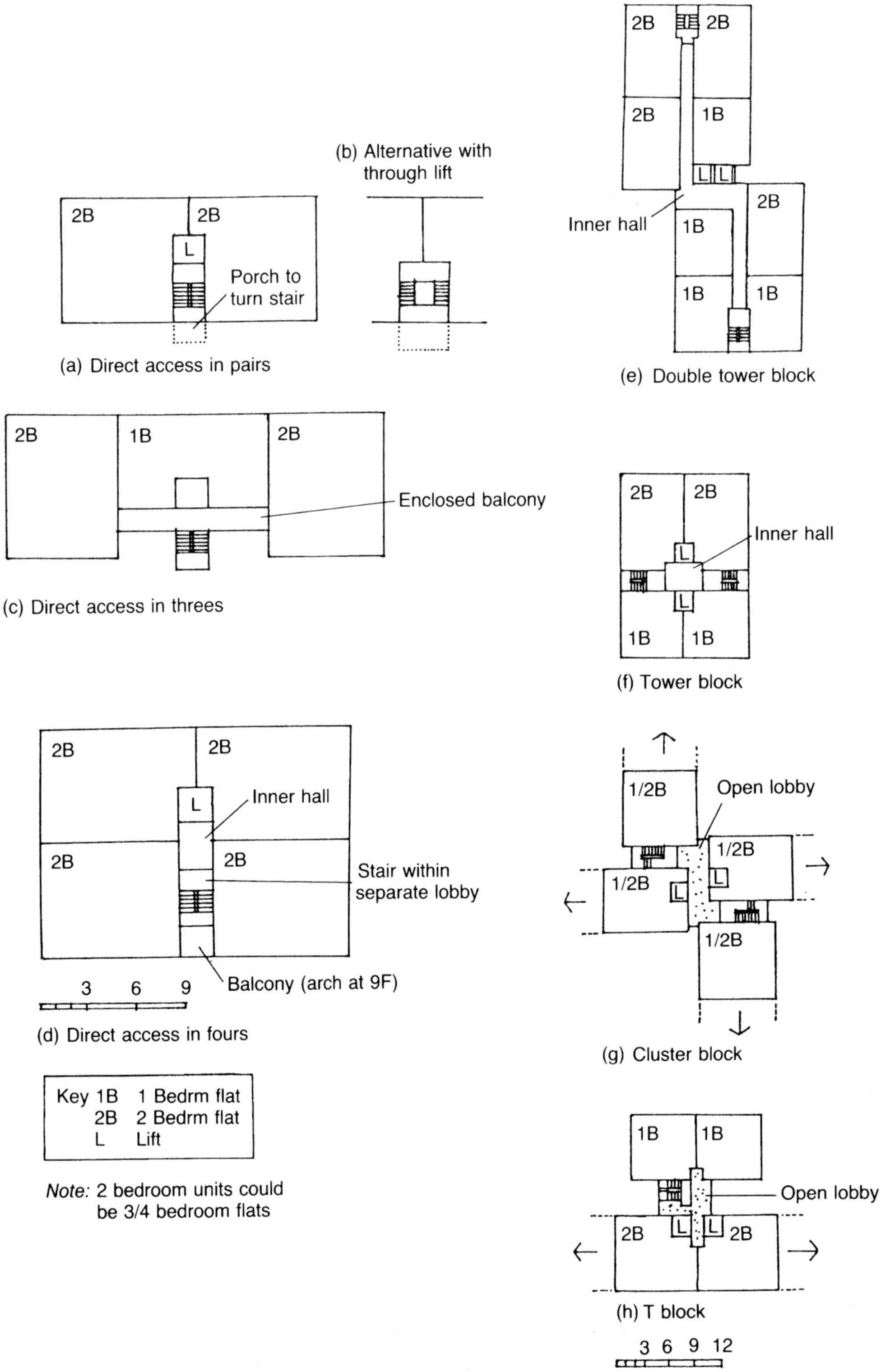

60 Stairs, Steps and Ramps

Figure 3.28 Generic analysis for access cores

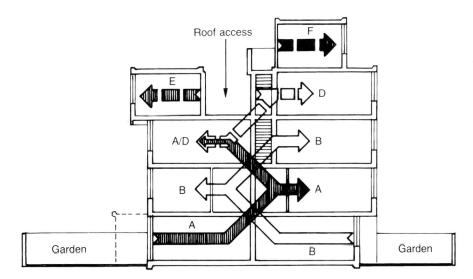

Figure 3.29 Scissors-type stairs at Lillington Gardens, 1963–71. (Darborne and Darke) Plan and Section. (See Figure 3.7 for detailed section through stairs)

well as individual dwellings is provided in Chapter 11.

Fire insurance precludes spiral forms so that direct flights, dog-legs or three-turn stairs are the most common forms. Lifts, in the case of a fire, are often reserved for fire fighters which implies that the lift cage is separate and no longer enclosed by three-turn stairs (as Figure 3.28*b*). Such enclosure was a popular economic form devised in the last century. Separating the lift from the stairs has the advantage that 'through lift' access is feasible, giving considerable benefit when planning for pram or wheelchair use.

The 'scissors plan', based on dog-leg stairs serving intermediate floors of maisonettes, came to prominence with the Lillington Gardens competition at Westminster, won by Darborne and Darke in 1963. The compact, interlocking form has the greatest economy when placed at right angles to a balcony or central passage. The dwellings are lined up as terraces of 'Universal Plans' based on narrow and wider modules (Figure 3.29). The compactness of the plans is explained by the high density requirement of 300 persons per acre and where the designers remodelled the familiar London house with a balcony disguised as a garden deck complete with planters and sprinkler system.

The disadvantage rests with the need to use steps to reach most 'servant' elements such as bathrooms or even kitchens from living areas. Not an ideal arrangement for the disabled!

3.3.3 Preferred tread-to-rise dimensions

A few words are needed on preferred tread-to-rise relationships. Ideally a common stair proportion should be used throughout a residential scheme, say 190 mm rise × 240 mm going for both common and private stairs.

The British Regulations are however very contrary and permit the steeper ratio of 220 mm × 242 mm going at a pitch of 42° within the private confines of a flat. Co-ordination of floor levels will clearly be a problem if differing risers are chosen since the upper floors of maisonettes may well have escape routes leading out to fire stairs (with 190 mm × 240 mm rise to tread). Common sense dictates a common riser of say 190 mm with treads perhaps adjusted for common and private use. In the case of scissors type maisonettes, where public and private steps are side by side, common proportions clearly have to be used throughout.

Research concerning ease of use, safety and comfort during ascent and descent has been carried out by various agencies. A notable British example was inaugurated by Peter Randall for the Research and Development Group of MOHLG back in the golden days for R and D in 1967. These ergonomic studies were based upon men and women testing twelve alternative conditions. A scientific stance was assumed with physiological tests of heartbeat, physical records taken with multiple exposure photographs and finally psychological tests, with the sample users asked to place differing staircases in order of ease, safety and comfort.

The age group chosen were between 62 and 78 since it was felt that this older range could be critical in studying the ease and safety of stairs in use. The pooled ratings are shown in Table 3.1.

The conclusions drawn from the study were as follows. First that within the rise and tread dimensions followed, an easier going made for greater safety, rather than an easier rise. A greater foothold was seen to be of prime importance. The second conclusion that risers might be increased to 215 mm or 229 mm seems hard to justify except in terms of modular floor to floor heights that were targets in the 1960s. Nothing transpired to change the awkward stairs preferred by government agencies and which are now enshrined in the standard stair offered by the timber trade with 200 mm × 223 mm as 'nominal'. A

Table 3.1. Riser-to-tread relationships in order of ease of use, safety and comfort

	Men		Women	
	Rise	Going	Rise	Going
1	178 mm (7") ×	291 mm (11 1/2")	178 mm (7") ×	266 mm (10 1/2")
2	178 mm (7") ×	241 mm (9 1/2")	178 mm (7") ×	291 mm (11 1/2")
3	178 mm (7") ×	266 mm (10 1/2")	203 mm (8") ×	266 mm (10 1/2")
4	203 mm (8") ×	266 mm (10 1/2")	203 mm (8") ×	291 mm (11 1/2")
5	203 mm (8") ×	291 mm (11 1/2")	178 mm (7") ×	241 mm (9 1/2")
6	178 mm (7") ×	215 mm (8 1/2")	203 mm (8") ×	241 mm (9 1/2")
7	203 mm (8") ×	241 mm (9 1/2")	178 mm (7") ×	215 mm (8 1/2")
8	229 mm (9") ×	291 mm (11 1/2")	203 mm (8") ×	215 mm (8 1/2")
9	229 mm (9") ×	266 mm (10 1/2")	229 mm (9") ×	291 mm (11 1/2")
10	203 mm (8") ×	215 mm (8 1/2")	229 mm (9") ×	266 mm (10 1/2")
11	229 mm (9") ×	241 mm (9 1/2")	229 mm (9") ×	241 mm (9 1/2")
12	229 mm (9") ×	241 mm (9 1/2")	229 mm (9") ×	215 mm (8 1/2")

retrospective appraisal of the former MOHLG deliberations can be made as follows. Most people prefer a shallower pitch than normal, the stated preferences for risers as shallow as 178 mm with goings as generous as 266 mm to 291 mm give comfortable steps – a point of detail met by carpenters' work in the eighteenth and nineteenth centuries but forgotten by the building industry today.

3.3.4 Elevational considerations

Reference to the generic forms in Figure 3.28 demonstrates the basic dilemma, namely to plan arrangements that permit maximum fenestration to the flat units. For example in Figure 3.28, plan 'a' would provide a two window elevation while plans 'g' and 'h' can provide three.

Internal staircases, if arranged to meet Fire Brigade needs, will present the easiest solution with top lighting provided for daytime requirements. Reference to Maison Clarté (Figure 8.5e) reveals the way central stairs designed with glass treads can permit a rooflight to filter down daylight through eight storeys. Ottoman domestic architecture often embraced a similar concept with minimal open balustraded stairs placed on the periphery of a well lit shaft (refer back to Figures 2.22a, b).

Stair shafts placed outside the building volume become towers whether lightly framed or faced in solid materials (refer back to Figures 3.25a, b). In long slab blocks, external towers can break up the building into identifiable elements (Figure 3.30a). It is a matter of scale, however, since the same concept applied to tall flats, say 12 or more storeys, increases the sense of megalomania (Figure 3.30b).

Stair and lift cores placed as nodal elements between flats have the advantage that shading will not occur and can enable a more subtle contrast to be achieved between the domestic and staircase fenestration (Figure 3.30c). The example in Figure 3.30c is taken from the work of Bruno Taut and has a refined detail whereby the stair window reveals are reduced by storey in contrast to the domestic pattern adjacent. 'T' or 'Y' shaped blocks give the opportunity for a fully glazed link between elements (Figure 3.30d).

Other treatments evolve from the profile of staircase enclosures brought across the façade. Both the Aalto designed dormitories at Cambridge, MA for MIT and the Crown Reach apartments by Lacey and Jobst share this feature on the accessible face of their designs. (Figures 3.30e, f).

Another version is the courtyard plan of Harvey Court, Cambridge, England by Sir Leslie Martin where the stairs climb parallel to the block. The underside of the ziggurat form permitting the flights to infill the colonnade can be expressed as part of the elevation (Figure 3.30g).

Very expressive forms were popular in association with balcony access flats, the modernist designs by Wells Coates capturing the contrasting ideals between the sleek horizontals of the balconies with spiral ribbons of concrete to the stair balustrades (Figure 3.30h).

Combining the entry arrangements to flats with refuse chutes is one of the more obnoxious ideas in British housing policy and particularly where the terminating chute and communal dustbin has to be situated close by the ground floor hallway. Basement servicing for refuse is commonplace in continental Europe and the USA and which explains why flat entries and stairs from abroad have been chosen for the final illustrations (Figures 3.30d and 3.31).

64　Stairs, Steps and Ramps

Figure 3.30 Staircase fenestration in flats
a Identification elements, Fasenenplatz, Berlin, 1984. (Gotfried Bohm)

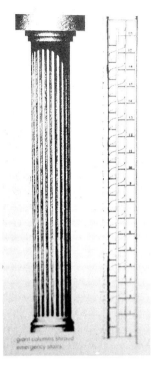

Figure 3.30b Megalomania: Les Espaces d'Araxes, Marne-la-Vallee, 1982. (Bofil)

Stairs, Steps and Ramps 65

Figure 3.30c Britz Siedlung, Berlin, 1931. (Bruno Taut)

Figure 3.30d Bellahoj, Copenhagen, 1954. (A. S. Dominia)

Figure 3.30e Baker House Dormitory, Cambridge MA, 1948. (Alvar Aalto)

Figure 3.30f Crown Reach, Millbank, London, 1981 (Lacey & Jobst)

Figure 3.30g Harvey Court, Cambridge, 1962. (Sir Leslie Martin and Colin St John Wilson)

Figure 3.30h Lawn Road Flats, 1933. (Wells Coates)

Figure 3.31 Attractive hallway and entry stairs to flats: Doldertal, Zurich, 1934. (Roth Brothers with Marcel Breuer)

References

[1] 'Generic Plans' is a term developed by the former GLC & Ministry of Housing. Both organizations published 'preferred plans' on the generic basis. Useful references are *Space in the Home* DOE HMSO and NBA (1969) *Metric House Shells*.

[2] Aalto's preference for the Centimetre Module was revealed in the author's conversation with Reima Pietila.

[3] Mrs Schroder back in 1972 (when the photo was taken) claimed that the steep winders kept her alert and fit. She was then in her early seventies.

[4] It is difficult to pin down the source of the 'London' house plan but John Summerson in *Georgian London* recognizes that Nicholas Barbon was the first speculator to extensively build this pattern of house after the Fire of London. The constructional method with brick envelope walls to a studwork interior being standard practice until 1914.

[5] The eighteenth-century landscape garden on St Annes Hill was created by Walpole, the focus being the semi circular portico recaptured by Raymond McGrath in the modern house in 1935–6, for details refer to Christopher Tunnard's book *Gardens in the Modern Landscape*, 1938.

[6] DOM-INO skeletal house was an original design patented by Le Corbusier and Max du Bois in 1914.

[7] The National Trust are due to take over the Goldfinger House and it will be interesting which phases in the family history will be exposed to view.

[8] The lucky few in Ostia or Rome (AD 100) enjoyed water borne drainage, namely those buildings built close by or below an aqueduct structure, *Everyday Life in Ancient Rome* by F. R. Cowell (refer to p. 22).

[9] Donald J. Olsen describes the mid-nineteenth-century flats of London in 'The Growth of Victorian London' Ch. 3 'Blocks of Flats' (pp. 114–18).

[10] The notes on balcony access stairs are taken from the Ministry of Health Housing Manual 1949, but by then the MOH were saying that under some circumstances balconies might be of any length.

[11] Refer to layouts for Unite d'Habitation Marseille (1945) in *Oeuvre Complete 1938–46*.

4 Commercial stairs

4.1 Core planning and standardized stairs

Lift and stair cores in commercial buildings can be grouped according to building use – hotels, offices, retail and special facilities (such as multi-purpose layouts that contain conference and exhibition space). Building Codes in Britain stipulate a minimal tread-to-rise relationship that varies according to the 'use' category.[1]

A wise designer ought to take the most generous proportion available in multi-use buildings so that a common denominator in stair risers are maintained throughout. This means that standardized stairs can be made and a modular approach adopted to vertical dimensions within the total building section (Figure 4.1).

Under British Codes, the 180 mm riser × 280 mm tread provides the statutory minimum stair proportion for all categories of use within a multi-purpose design. This ratio however, does not match the comfort of the easy going 150 mm × 300 mm steps that served a similar role in the days of Sir Christopher Wren and which related to a brick module of 75 mm.

Today, commercial considerations play a dominant role in demoting stairs to a secondary place whilst lifts or escalators assume greater importance (Figure 4.2) Valuation surveyors run their rules over staircase dimensions to establish that minimal criteria have been applied since core plans do not count within the lettable area. The design process with hotels or offices must take this into account and every effort made to reduce wastage in core planning. In fact larger

Table 4.1 Extract from categories under the Building Regulations in England and Wales (1991)

Category	Maximum rise	Minimum going	Maximum pitch
Private stairways	220 mm	220 mm	42°
Institutional and assembly buildings	180 mm	280 mm	35°
Institutional and assembly buildings with an area less than 100 m^2	180 mm	250 mm	35°
Other buildings	190 mm	250 mm	35°

In previous regulations there was a special category called 'common stairs, related to flats' where the 190 × 240 ratio was permitted, clearly extension of such buildings in future could raise problems with non-matching goings.

Stairs, Steps and Ramps 69

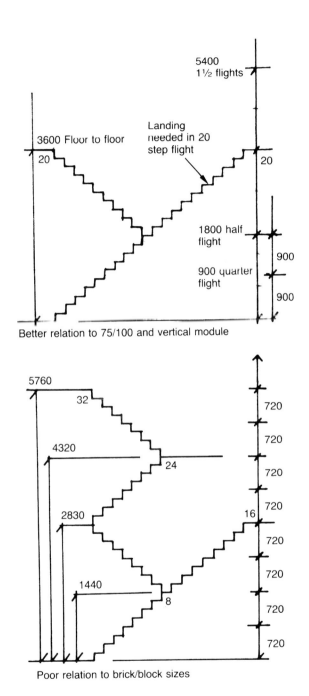

Figure 4.1 Modular heights for steps

Figure 4.2 Lifts and escalators assuming greater importance – interior of Lloyds, London, 1986. (Richard Rogers and Partners)

scale layouts are worth developing at an early stage to establish options for floor to floor heights in relation to the size of stair shafts that have to be accommodated. A well 250 mm wide not only looks attractive but can absorb two risers at either end which in turn allows a 720 mm variation in storey height without increasing the overall size of the shaft (Figure 4.3). Floor areas and occupancy rates provide the traffic figures for lifts and stairs. Sizes for the former can be taken from installers' catalogues whilst the latter has to be determined by consulting Building Codes, or else established with the Fire Brigade or licencing authority; an outline is given in Chapter 11. The key dimensions relate to the maximum numbers of steps within an unbroken flight and the number and

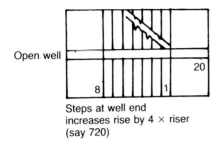

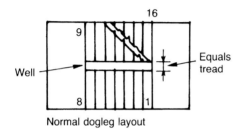

Figure 4.3 Steps at well end

width of stairs needed for escape purposes.

Compactness with escape stairs can be achieved by stacking two sets of flights one above the other with a solid reinforced concrete wall separation (Figure 4.4a). A dwarf reinforced concrete fender wall, say 100 mm to 150 mm thick, to infil the well takes up less space than a conventional metal balustrade, particularly with those designs which have swept handrailing at the well ends (Figure 4.4b). Non-continuous handrails or stepped forms save space (refer also back to Figures 3.22a, b). A half step relationship across the well in dog-leg stairs will ease the geometry of handrailing and only add half a step in the length of the flight (Figure 4.4c). The constructional Chapters 8 and 9 detail this arrangement.

The completion of the preparatory work can provide a matrix by establishing the repetitive upper floor plans whether hotel, office or retail. It is common practice to set out the main elements of the upper storeys in order to develop the 'footprint' of the building at ground level (Figures 4.5a, b, c, d). The arrangement of the Criterion layout by Renton Howard Wood Levin is of considerable interest – the key rests with the 'L' shaped section of the atrium space with foyer and escalators at street level leading to a finely made curved glazed shaft containing wall climber lifts. Conventional cores with escape stairs, lifts and lavatories exist in the adjacent space but new technology with fire resistant glazing has transformed the concept.

The dichotomy that plagues core design rests with the attractions needed at entry level in comparison with the minimal qualities more appropriate to the lettable space on the upper floors.

The simplest solution is to unwind the principal flight into a more generous geometry at the entrance hall. A typical example is where a dog-leg design turns into an open well arrangement. Present day developments with fire resisting glazing can recapture transparency (refer to Figures 4.5d, 4.6) instead of stair shafts being hidden away in walled off enclosures. The internal core of the offices at Milton Gate designed by Denys Lasdun, Peter Softley and Associates makes an equally dramatic statement with the circular shaft containing lifts and stairs being connected by bridges to the glazed fenestration of the surrounding offices.

Accommodation stairs are by constrast seen as assets in commercial buildings and often become the focus of attention within prime circulation areas. Such arrangements are additional to the mandatory means of escape and can therefore be designed with curving or tapered forms and of course be free of fire enclosures (Figure 4.7). (Refer to Section 4.2 for details.)

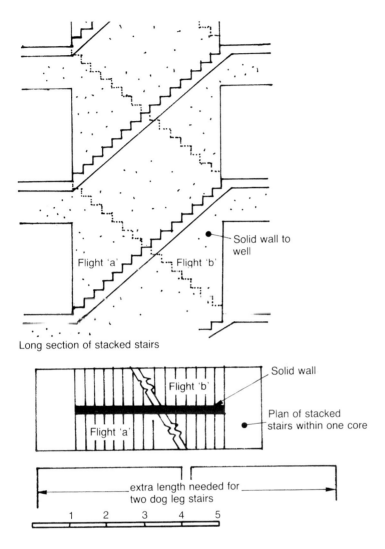

Figure 4.4a Solid wells. a Stacked escape stairs

4.1.1 Office core locations

The salient dimensions relate to the maximum escape run from the furthest corner of the working area to the safety of the enclosed stairs. Protected lobbies or passages can extend these lengths. The basic arrangement is given in Figure 4.8a with guidance on limiting factors in the UK and abroad (Figures 4.8b and c).

The core locations depend upon the floor configuration adopted. For high rise one has to look at slab block versus tower. For medium rise further comparisons need to be made with atrium and courtyard forms as well as geometries involving L, T, U, X and Y shaped buildings. Figures 4.9 and 4.10, provides a diagrammatic guide with references to typical designs. The obvious criteria from the economic view point are core sizes that can be expressed as the smallest proportion of the rental and usable floors. For users there is another critical factor, namely the largest areas that can be accommodated on any single floor despite

72 Stairs, Steps and Ramps

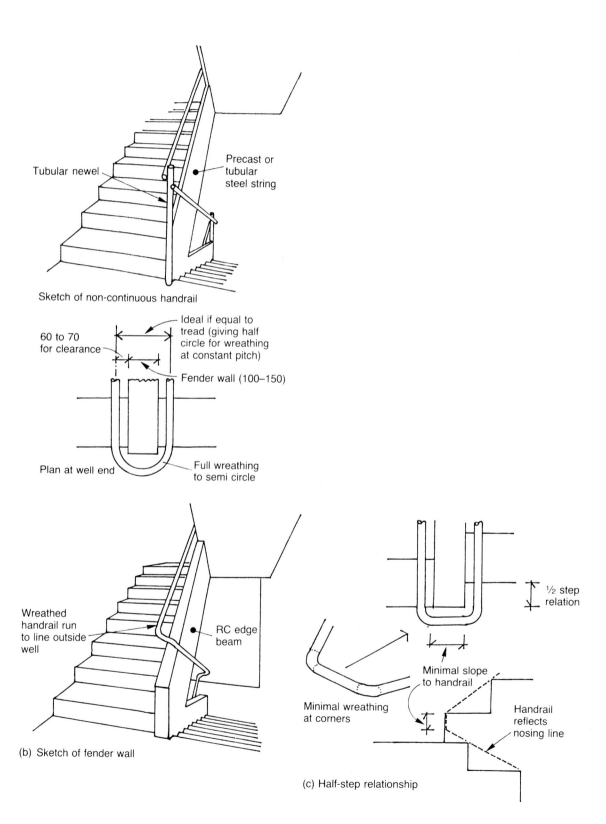

Figure 4.4b Solid wells   c Half-step relationship

Stairs, Steps and Ramps 73

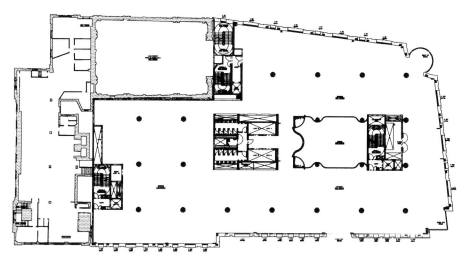

Figure 4.5 Matrix design for cores and footprint at street level, example from the 'Criterion' Piccadilly, 1992. (Renton Howard Wood Levin)
a Street level plan

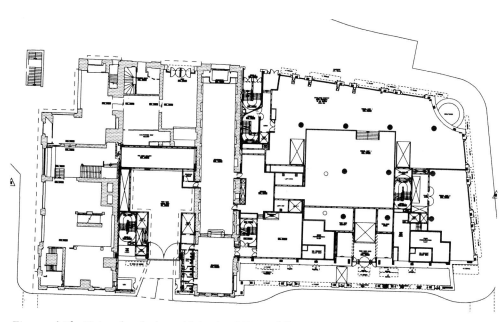

Figure 4.5b Upper level plan, 'Criterion' Piccadilly

74 Stairs, Steps and Ramps

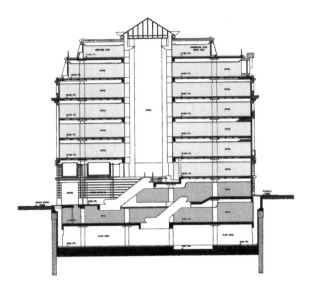

Figure 4.5c Section of 'L'-shaped profile to atrium, 'Criterion' Piccadilly

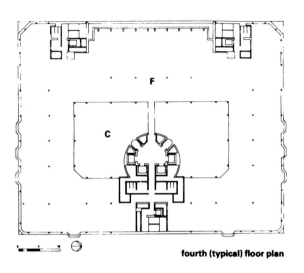

Figure 4.6 Clear fire screen glazing at Milton Gate, City of London, 1992. (Denys Lasdun, Peter Softley and Associates) Key plan

Figure 4.5d View within glazed atrium with escalators and wall climber lifts, 'Criterion' Piccadilly

Figure 4.7 Accommodation stairs, Legends Club, London, 1990. (Eva Jiricna)

Stairs, Steps and Ramps 75

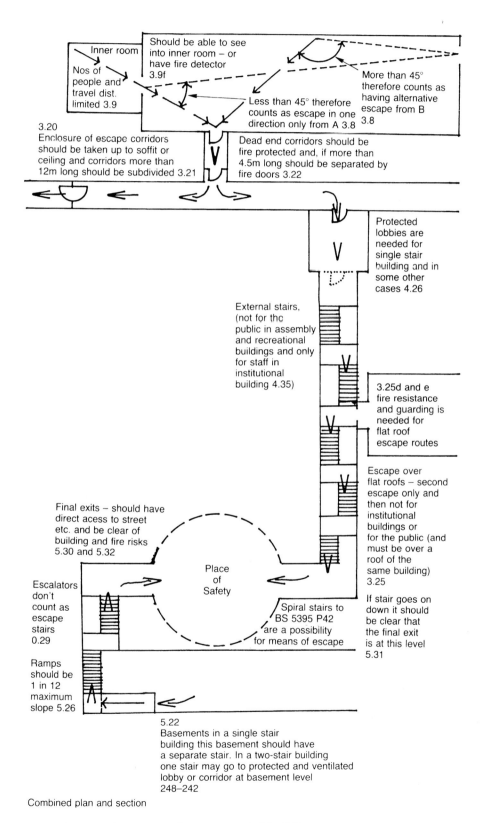

Combined plan and section

Figure 4.8 Core limitations for commercial buildings
a Limitations imposed by National Building Regulations

76 Stairs, Steps and Ramps

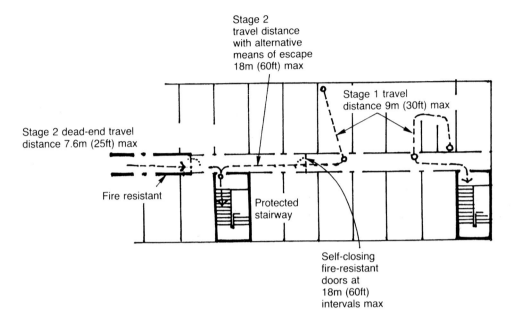

Figure 4.8b Typical specified travel distances to stairways in a hotel building given by international planning guides

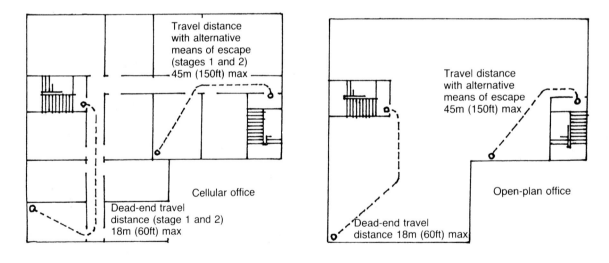

Figure 4.8c Typical guidance for offices published in West Germany from the early 1980s

breaks incurred through lift and stair locations. The studies prepared by Sir Leslie Martin in the 1960s for the rebuilding of Whitehall demonstrated that interlocking courtyards would prove more adaptable than isolated slab or tower blocks in medium rise buildings. The present interest with atria based forms has produced deep floor plans of extended 'race track' layout with cores distributed along the edges as at No. 1 Finsbury Avenue, Figure 4.10e, and for IBM at Bedfont Lake, described in Case studies 14.12 and 14.13. Triple bay depths have the advan-

Stairs, Steps and Ramps 77

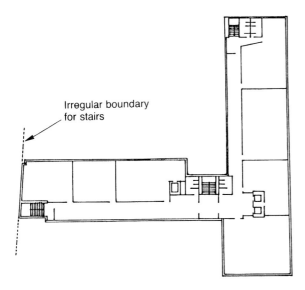

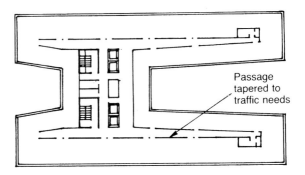

Figure 4.9d Double zone layout using H or U form, typical pattern to devise window walls to all sides of offices. Variations include V, X and Y forms around core zones

*Figure 4.9 Core arrangements for offices
a Single zone layout: main stair and lifts at junction of two wings, with secondary stairs at ends of corridors*

tage of layering plans so that the central zone can accommodate cores and servant spaces with the daylit outer zone used for prime activity. Race track plans of this pattern are used for highly serviced office buildings (Figure 4.10). They also apply to educational facilities, hospitals and

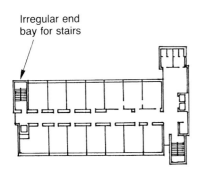

Figure 4.9b Double zone layout: spine corridor form with main stair, lifts and service area to one end and secondary stair and lift at opposite end

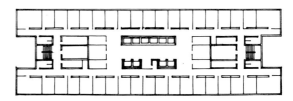

*Figure 4.10 Multiple bay plans.
a Triple zone layout with parallel corridors and central core (Phönix-Rheinrohr AG Germany Competition 1955)*

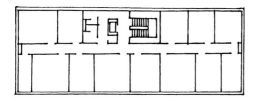

Figure 4.9c Double zone layout with central core: typical single staircase design for low-rise could however form a pattern for development with equal sized cores spaced at intervals in length of block

Figure 4.10b Triple zone with 'race track' plan enclosing central core (Phönix-Rheinrohr AG Germany Competition 1955)

78 Stairs, Steps and Ramps

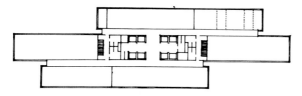

Figure 4.10c Staggered triple zone layout (Phönix-Rheinrohr AG Germany 1957, scheme completed). (Helmut Hentrich and Hubert Petschnigg)

laboratories where servicing is centrally located.

More extreme separation of the elements can be seen at Lloyds HQ, London and with the Hongkong and Shanghai Bank where external towers provide lift and stair cores (Figure 4.11a, b). These 'servant' features are placed outside the uninterrupted volume of atria and related offices. Case studies are presented in Chapters 12 and 13 with the escalators for both buildings illustrated in detail.

4.1.2 New directions

The reconstruction of the former Financial Times building by Michael Hopkins and Partners has achieved a unique subdivision with the central atrium and lift cores placed at the maximium permitted remove from the compartmented escape stairs and remodelled

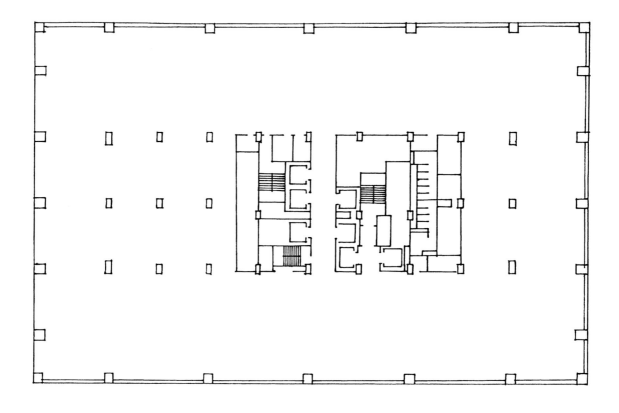

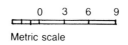

Metric scale

Figure 4.10d Open layout for high rise tower, John Hancock Centre, Chicago, Illinois, 1969. (SOM)

Stairs, Steps and Ramps 79

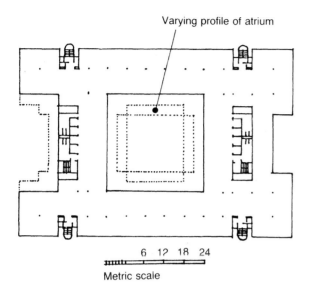

Figure 4.10a Deep plan zones around atria spaces. The dimensions permit roughly consistent depth of office from fenestration. No. 1 Finsbury Avenue, City of London, 1985. (Arup Associates) (further details given in case study 14.12)

Figure 4.11b Hongkong and Shanghai Bank, 1986. (Foster Associates)

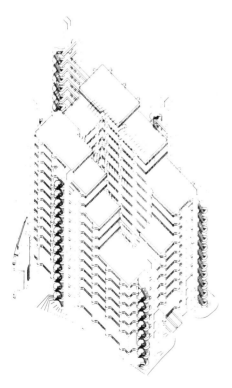

*Figure 4.11 External cores
a Lloyds HQ, London, 1986. (Richard Rogers and Partners)*

areas within the retained wings of the former layout (Figure 4.12a). Accommodation stairs are sited either side of the new open offices and permit direct connection between the floors throughout the eight storeys. The perforation of compartment floors is allowed due to the sprinkler system installed throughout the common office accommodation and to the fact that four new escape stairs are strategically designed at either end of the retained wings. A further benefit of the layout with outward escape from the atrium lift lobby is relaxation of fire enclosure to the main vertical circulation. The steelwork is uncased and the fenestration openable

80 Stairs, Steps and Ramps

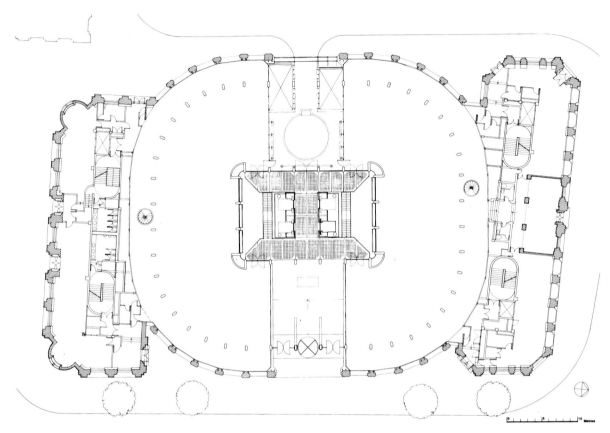

Figure 4.12 Bracken House, City of London, 1992. (Michael Hopkins and Partners)
a Plan at street level

between working areas and atrium and hallways. It is another version of the Bradbury Building in LA (Figure 4.31) but more compact and relevant to todays working needs. The role played by the vertical circulation elements here indicate a change of emphasis, with lifts now having primary importance; they are a feature in the atria – as wall climbers – while structural support is given by steel plate towers (Figure 4.12c). The circular accommodation stairs are the next most used element whilst the wings contain the four new escape routes (Figure 4.12b) and ancillary lifts. In case of fire, the Brigade teams will tackle the problem from the wings where allocated lifts and dry risers are sited either side of the open plan area. The layout combining new and old elements on the Financial Times site has produced a new concept of conservation and one that shows the most interesting combination of technology in arranging lifts and stairs today.

The other reconstruction indicating changes in attitude to fire safety is the story of the Economist Building, with respect to the original core plan (*circa* 1964) and the remodelling undertaken in 1992. The building, like Bracken House, falls under the Section 20 legislation that applies to 'over cube' premises in Inner London, – a definition is given in Chapter 11, Section 4.

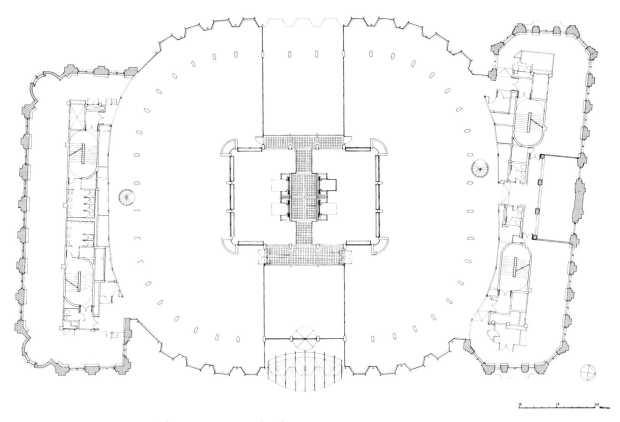

Figure 4.12b Atria and lift cores at upper level

The initial designs had to comply with the former requirements of the London Fire Brigade to have facilities for climbing directly onto escape stairs in case of lift failure. Such demands, with internally sited stairs, meant two open air shafts within the core. Each shaft had a rung ladder accessible to windows on both escape route landings (Figure 4.13a). It was a solution favoured in old Parisian apartment blocks. The changes brought about by 'positive' air pressure devises for stair and lobbies have revolutionized the layout of inner cores. At the Economist Building it enabled the service area to be enlarged to contain a kitchen as well as male and female lavatories at every floor as opposed to alternate levels (Figure 4.13b).

4.1.3 Hotel core details

Lift and stair locations in hotels depend upon permitted lengths of protected corridors serving bedroom accommodation.[2] The significant lift and main stair locations are based on the building's footprint at entry level. This is not necessarily related to the street but to the upper part of the podium. The Renaissance Center Detroit (Figures 4.14a, b, c) is a prime example. Here, escalators, lifts and escape stairs serve the podium with a transfer to the main core within the entry to the hotel,

82 Stairs, Steps and Ramps

Figure 4.12c Steel plate towers to support wall climber lifts, Bracken House. (Courtesy of Alan Delaney)

to afford greater security. At the author's visit there was an electro-magnetic device to search guests for weapons in the hotel foyer.

Many Portman designed hotels[3] dramatize the entry zone with a vast atria surrounded by galleries served by wall crawler lifts from top to bottom of the space (Figure 4.14d). A modest variation is the Stirling Hotel, London Airport where the lifts form a vertical feature within a generous atria whilst the fire stairs are lost within the bedroom wings to accord with British regulations (Figure 4.15). Another version in the UK is the Gatwick Hilton where the courtyard form permits the stairs and lifts to fall within the internal corners, these in turn relating to the periphery servicing of the public spaces. The main entry stairs and escalator foyer have been generously proportioned and look extremely well in the context of the first floor (Figure 4.16a, b).

Lift cores are clearly important but it is noticeable at Gatwick that the stairs make

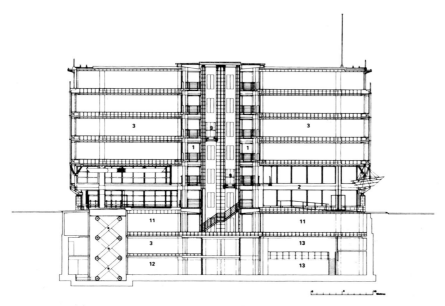

Figure 4.12d Cross section

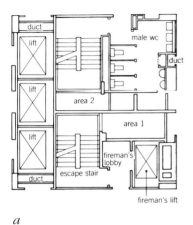
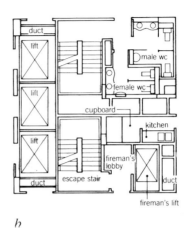

Figure 4.13 The Economist Building
a Original core, 1964. (Alison and Peter Smithson in association with Maurice Bebb)
b Reconstructed core, 1990. (SOM)

the public spaces work together. The final examples demonstrate the key roles of staircases in older buildings. Firstly, there is the Ritz Hotel, London designed by Mewès and Davis in the early 1900s. The 'U' shaped plan placed the main facade to Piccadilly with return wings to Green Park and to the main entry off Arlington Street. The sequence of public spaces runs east to west towards Green Park, with a large open well stair placed as an eye catcher turned towards the principal suites at first floor level (Figure 4.17). Lift lobbies and secondary stairs occur off the main axis but are subservient to the grand stairs first glimpsed off the main entry.

By contrast, at the Lanesborough Hotel – a conversion from the former St Georges Hospital of 1827 – the new internal circulation has been resited to emphasize lifts and vestibules, while the former stair hall now has a minor function and relates simply to the basement dining area. The scale of the remodelled hallway is a reminder of the grandeur of stairs before lifts took over in hotel planning (Figure 4.18)

Figure 4.14 Portman designed hotels, Renaissance Centre, Detroit, 1977.
a General view of multi-storey mall below hotel towers

84 Stairs, Steps and Ramps

Figure 4.14b Restaurants at lower level with stairs and bridges above, Renaissance Centre

Figure 4.14c Stair complexity equal to Piranesi

Stairs, Steps and Ramps 85

Figure 4.14d Hyatt Regency, Detroit, 1970s. Wall climber lifts playing a dramatic role within heavily shaped structure

4.1.4 Escalators, lifts and stairs in major spaces for retail and related buildings

The nineteenth-century department store and the great exhibition halls of that period are twin sources of a whole range of building types. It is worth recalling the layouts invented by Schinkel for the first multi-storey store, which developed 'master and servant' spaces very much in the manner used by Richard Rogers 150

Figure 4.15b View of central lifts, (Courtesy of Michael Bryant)

years later. Schinkel's stairs are tucked behind the selling floor that envelop the atrium-like interiors. One has to remember that the powered lift had not been invented at this time. The Schinkel project

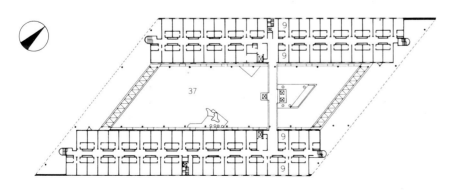

Level 2 - 5

Figure 4.15 Central lifts, Stirling Hotel, London Airport, 1990. (Michael Manser Associates) a Key plan

86 Stairs, Steps and Ramps

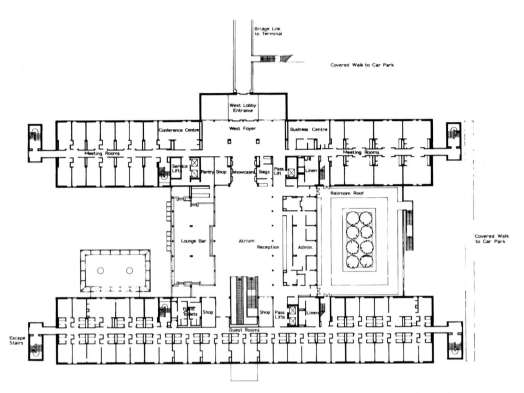

Figure 4.16 Gatwick Hilton, Surrey, 1980. (YRM Architects) (Courtesy of YRM)
a Key plan at first floor

Figure 4.16b Stair design within foyer

Stairs, Steps and Ramps 87

Figure 4.17 Grand staircase, Ritz Hotel, London, 1905. (Mewès and Davis)

called the Bazaar was proposed for the Unter den Linden, Berlin in 1827 and sought to replicate the ambiance of the Palais Royal arcaded shops built in the eighteenth century.

By comparison Paxton's Crystal Palace had periphery galleries and servant stairs (Figure 4.19) but also advanced the visual delights by placing double spiral staircases at key positions in the main halls to persuade the public to explore the upper levels (Figure 4.20). The curving forms were positioned in such a way that a panoramic view of all the exhibits could be obtained, a device used in present day shopping malls (refer to Figure 4.25) where stair geometry, particularly landings and the orientation of escalators, is designed to maximize the visual exploration of the retail area.

Figure 4.18 Remodelled stairs and hall, the Lanesborough Hotel, Hyde Park Corner, London, 1992. (Fitzroy Robinson)

Figure 4.19 Periphery galleries around atria spaces, Crystal Palace, 1851. (Designer Joseph Paxton)

Schinkel's Berlin store was never constructed but the design was certainly an inspiration for what had been constructed by the early twentieth century. In the simplest derivation galleried halls were provided, resembling covered market spaces, the GUM store in Moscow being typical. Whiteleys Store in Bayswater, London was certainly inspired by Paxton's Crystal Palace, young William Whiteley having been to the 1851 exhibition building in Hyde Park. He decided to transform his shops into shopping halls, although the transfer of his idea into reality at Queensway took 60 years. The final version with halls and balconies was constructed 1908–12 to the designs of Belcher and Joass with a splendid multiple stair in the spirit of the Crystal Palace double flight.

There are however two key department stores where highly imaginative stair design played a leading role. Firstly Les Galeries Lafayette, Paris where sinuous curves of the French Art Nouveau weld the steps and balconies to the upper floors in one simple composition (Figure 4.21). This is one of the most captivating examples of early department stores to survive. The effect is reinforced by Louis Majorelle's decorative line to the balustrading and handrailing which still decorates the interior space.

The second creation is the Samaritaine store (1901–10) by Franz Jourdain where he developed the constructional ideas with glass treads to the stairs and hidden lighting within glass paved balconies. The forms for treads and balustrades emulated lily leaves and tendrils. Regrettably most of the original work was destroyed by the modernization with Henri Sauvage in the late 1920s. The work by Hector Guinard in the same style has been better

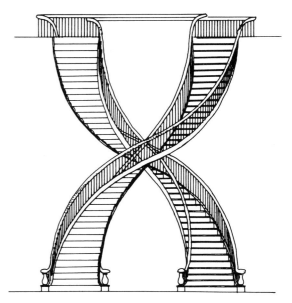

Figure 4.20 Panoramic stairs, Crystal Palace, 1851. Joseph Paxton. Designer of stair Lesley Banks

preserved in many of the entrance stairs to the Paris Metro (Figure 4.22).

The dictates of fire engineering led designers to look towards iron and steel construction, fully exposed in early examples but encased in concrete by the 1920s. The significant designs within the Modern Movement stem from the Schocken Stores and other commercial buildings designed by Erich Mendelsohn. The Cafe Astor layout given in Figure 4.23 illustrates the key role that Mendelsohn gave to the dynamics of movement within the plan. Ten cylindrical or curved stairs connect the various levels of restaurants, public spaces and vestibules.

Other seminal work stems from Prague where various covered arcades or multipurpose atria were created to serve as offices, shops and showrooms. The best known is the galleried stair hall of the Electricity Board, completed in 1935, designed by Adolf Bens and Josef Kriz

Figure 4.21 Sinuous stairs and balconies, Les Galeries Layfayette, Paris, early 1900s. (Stair designer Louis Majorelle)

Figure 4.22 Forms emulating lily leaves and tendrils, entry to Metro, Paris, 1900. (Hector Guimard)

(Figure 4.24). The construction is reinforced concrete with a top lit roof formed by glass blocks set in concrete ribs. The triple return flight is lit dramatically from the landing windows. Lift clusters are placed in lobbies left and right of the main hall.

In Britain the first taste of pre-war modernism in store interiors can be seen with the freely arranged floors and sweeping escalators within the John Lewis Store, Oxford Street, of 1939.[4]

Today's store or shopping mall deploys escalators, lifts and stairs as 'accommodation' routes to the key levels within the cavernous interiors. Ideally the public should be able to explore the sequence of spaces in a logical manner

90 Stairs, Steps and Ramps

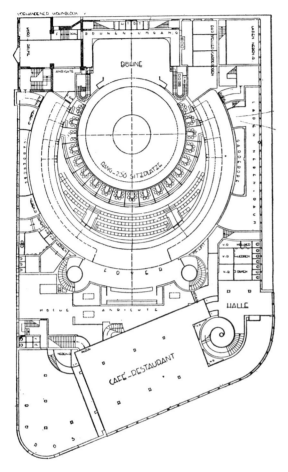

Figure 4.23 Mendelsohn and the dynamics of movement, Cafe Astor, Berlin, 1927

Figure 4.24 Pre-eminence of stairs, Electricity Board offices and showrooms, Prague, 1935. (Adolf Bens and Josef Kriz)

(Figure 25a, b). The illustration reveals the enticing way escalators or linkways are placed as nodes within a shopping mall to attract movement between floors. A logical flow both inwards and outwards from stores will be endorsed by the Fire Brigade but is thought to be less important by many shopkeepers. Their brief 'to entice the public in' is often more important than finding a convenient circulation back to the shop entrance. An exemplary case for clarity of movement via escalators is the recent John Lewis Store in Kingston-upon-Thames, the naturally lit space is bridged by an elegant array of moving stairs from floor to floor. Graphics exist hardly at all as they are not needed within the sales areas other than to locate fire

Figure 4.25 Covered malls.
a Shopping Mall, Malmo 1960s

Figure 4.25b Node locations with well-placed stairs and linkways, Georgetown Mall, Washington DC, 1980s

escapes (Figure 4.26). The success of the store has meant that extra escalators have been added to improve circulation.

The use of mechanical means of movement has transformed retailing design into a vertical pattern of developments, with banks of lifts and stairs set out as the key viewing point for the store interior. The concept has been transferred to multiple use buildings which might embrace hotels, offices and shops within a vertical atrium. Water Tower Place, Chicago has the drama of escalators and glazed lifts which serve the multi-storey mall, and a similar device dominates the Trump Tower in New York City. Both places have elevated moving stairs and exciting lifts into a new art form (Figure 4.27). The success rests upon the innate curiosity of people to ascend and move to brighter elements in the interior. Vertigo is combated by tinted glass balustrades, ledges and high level handrailing. It should also be explained that safety is also ensured in the Chicago location by the continual patrolling of armed guards.

4.2 Accommodation stairs

Accommodation stairs are defined as additional or amenity stairs in excess of means of escape provisions. In reality accommodation routes often form the most conspicuous features within commercial interiors. The key role is to link the crucial zones of the building, not only with moving people but with the creation of a visual inter-connection between the storeys. Portman's hotel designs are obvious examples where escalators, lifts and stairs are all employed to full theatrical effort within a vast atrium.

The benefits occur with more modest schemes, particularly in two or three storey buildings where a stair well allows the interior to be open to view. Eva Jiricna's well known designs[5] often relate to stores where movement through the various spaces are celebrated by the most imaginative flights of steps – Josephs store in Sloane Street is the best known example (Figure 4.28a, b). The use of glass and open framing gives the least interruption to the eye. Curving forms can be used to advantage to form an open cylinder within the plan. An imaginative play on this theme occurs within the Geschaftshaus, Vienna, designed by Hans Hollein. Here, the ascending geometry is

Figure 4.26 Interior of John Lewis Store, Kingston upon Thames, 1990. (Ahrends Burton and Koralek) (Courtesy of Chris Gascoigne)

created to afford views to each corner of the complex (Figure 4.29). The circular landings or stairs in differing widths are placed in differing planes to improve the upward perspective, encased lifts are placed alongside for functional servicing floor by floor while the centre piece is a veritable 'house of stairs' providing as much enjoyment as the Stiegenhaus in Bruhl.

The adjustment of stair widths is a feature used in nineteenth-century hotels where the width reduces with ascent, thus providing a tapering well and increasing the spacious quality when looking up the central well. This idea works most effectively with three turn designs within a squarish geometry. (Figure 4.30).

Original ideas were generated in the late nineteenth century as new building types were developed, one of the most inspirational atria interiors was the open metal stairs and visible lift shafts used by George H. Wyman in the Bradbury Building, Los Angeles (Figure 4.31). Another form of technical bravado is the Hallidie Building, San Fransisco (Figure 4.32) where the architect devised the escape stair as the principal feature,

Stairs, Steps and Ramps 93

Figure 4.27 New Art Form for Escalators and Lifts
a Water Tower Place, Chicago, 1976. (Loebl, Schlossman, Bennett and Dart)

Figure 4.27b Another view Water Tower Place

94   Stairs, Steps and Ramps

Figure 4.27c Trump Tower, New York, 1989

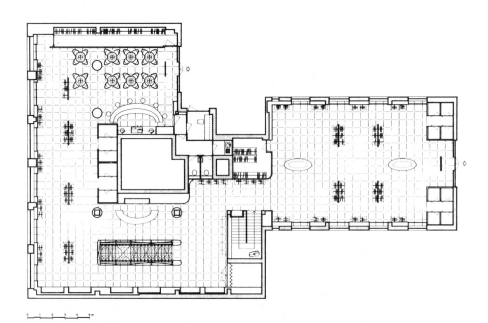

Figure 4.28 Accommodation stairs by Eva Jiricna
a Layout plan Joseph store, Sloane Street, London (see Figure 8.7a-c for details)

Stairs, Steps and Ramps 95

Figure 4.28b Interior space, Joseph store

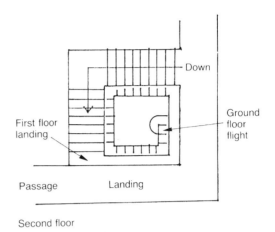

Figure 4.30 Increasing well size

Figure 4.29 Curving forms, Gerschaftshaus, Vienna, 1990. (Hans Hollein)

96 Stairs, Steps and Ramps

Figure 4.31 Bradbury building, Los Angeles, 1893. (George M. Wyman)

Stairs, Steps and Ramps

Figure 4.32 Hallidie Building, San Fransisco, 1917. (Willis Polk)

placed externally to the fully glazed facade, an idea seen again in the gable ends of the Centre Pompidou.

References

[1] Refer to a useful guide entitled *The Building Regulations* by Powell Smith and Billington (10th edn.), 1995. Note that different regulations apply to Scotland.

[2] Most major hotel groups produce design guides concerned with space standards, the sections related to protected cores and stairs being of greatest value.

[3] Portmans remarkable oeuvre is best seen in a book entitled *The Architect as Developer* by Portman and Barnett, 1976.

[4] The Lewis Partnership employed Slater and Moberly as project architects for both the Oxford Street store (1938–39) and for the Mendelsohnian Peter Jones store, Sloane Square (1935–37), the latter design however had the inspired benefit of a gifted designer William Crabtree as well as Professor Charles Reilly as consultant.

[5] For Eva Jiricna designs refer to *Eva Jiricna Designs in Exile* by Martin Pawley (1990) published by Fourth Estate and Wordsearch.

5 Civic and public stairs

5.1 General points

Civic architecture, unlike commercial development, often has a ceremonial or formal response to access and movement. This can be expressed in a sequence that encompasses the external space, the loggia or portals, the entry hall and thence to the primary, secondary and subsequent elements of the composition. Slight changes of level or steps can order the circulation whilst the design of major stairways can reinforce the various categories of importance or status of elements within the building layout.

These parameters apply historically to the Palace of Westminster (Figure 5.1*a*) or, today to the new parliament building in Bonn (Figures 5.1*b*, *c*, *d*). Both layouts are shown in outline to highlight the importance of stairs and the way these feature in the formal working of the plans.[1] The Palace of Westminster uses stairs to elevate the significance of the principal accommodation, which is arranged at first floor level where the formal axes connecting the main elements are to be found. The reasons are linked

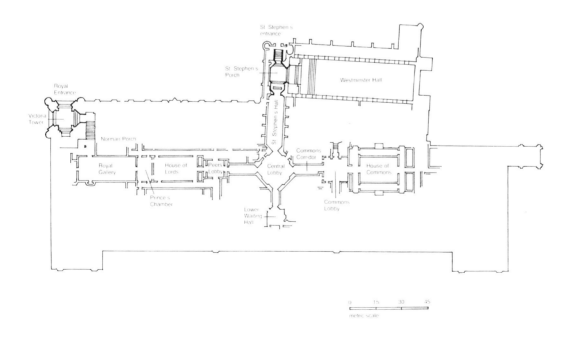

Figure 5.1 Common parameters showing the importance of stair locations within formal plans
a Outline of the Palace of Westminster

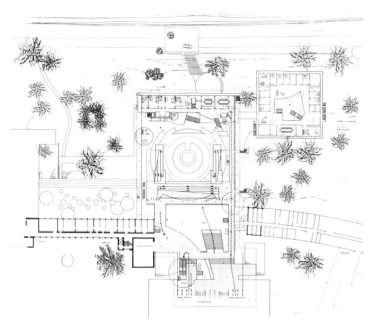

Figure 5.1b New Parliament Buildings, Bonn, Germany 1981. (Behnisch and Partners) Entrance level plan

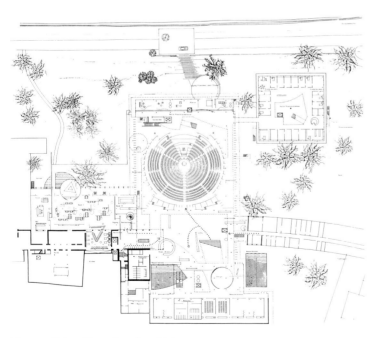

Figure 5.1c Chamber level plan

Figure 5.1d View from entry hall towards entrance, New Parliament Buildings, Bonn

to security but primarily derive from the Victorian fear of flooding from the Thames. The plan positions the House of Commons and the House of Lords on a long east–west axis, parallel to the river and forming the matrix for the corridor system connecting the principal parts of the Palace. The Royal approach is made via formal stairs signified by the Victoria Tower whilst public access takes place alongside Westminster Hall, terminating in the central lobby, which is the crossroads between the domains of the Commons and the Peers.

The provision of functional stairs connecting to other floors is tucked away as a secondary component throughout the Palace.

By contrast the new parliament complex in Bonn promotes a totally relaxed relationship between the public and parliamentary domain. First, the siting has been so arranged that the whole composition steps down towards the Rhine ensuring a better relationship between the interior and the natural landscape. The downward approach via splayed stairs through an open foyer coupled with the glazed chamber itself permits a visual connection between the public areas and the political arena. The concept is enhanced by balconies or galleries which bridge the foyer, enabling sight lines to be maintained throughout the spacious public interior. Lifts and conventional stairs provide direct visible connections between floors. Any subdivisions of spaces or restrictions on accessibility are due to security counters – bullet proof glazing being used at critical boundary zones to improve the safety aspect. A stepped descent is unusual for major public spaces but the advantage at Bonn is the ability to make the whole composition into a visual totality.

It is not possible to categorize circulation patterns in civic buildings with the ease with which it can be done for office blocks. However a broad subdivision can be made between plans using central space and periphery stairs – these are now demonstrated by significant examples, from the past and the present.

5.2 Central space stairs

The Augustusburg at Brühl has a magnificent double return stair as a centre-piece, the flights are not in fact, equal, to give a greater perspective to the upper ceiling. The *piano nobile* embraces the superbly scaled flights from both directions (Figures 5.2a, b). The open loggia at ground floor forms the carriage entrance with a sequence through the various

Stairs, Steps and Ramps 101

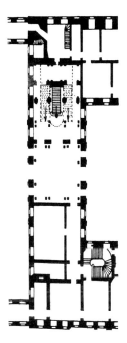

Figure 5.2 Baroque Palace Stairs. The Augustusburg, Brühl, Rheinland, 1743–48. (J Balthasar Neumann)
a *Outline plan of stairhall. (From Schuber, F.,* Treppen, *Hoffman Verlag, 1949)*

c *Residenz, Wurzburg, Germany, 1723–43. (J Balthasar Neumann) Use of return flight to turn axis back to central space*

b *View of return flights within the central hall. (Courtesy of Steinhoff)*

courts, while the secondary openings lead to offices and service stairs. Neumann explored similar ideas in the entry hall at the Residenz in Würzburg, where the greater length of the 'return' flights enables the axis to be turned towards the central space (Figure 5.2c). It is a pattern used in many custom made nineteenth-century art galleries, where first floor gallery space requires a formal approach.

One of the most interesting derivations is the design made by Semper for the Kunsthistorisches Museum, Vienna (Figures 5.3a, b, c). This is archetypical of the central space stair applied to a large public building and certainly equals palace planning of the baroque era. The double courtyard arrangement has the main entry placed at right angles through the central wing, the spatial effect is enhanced by spatial volumes that penetrate to the upper storeys. The vestibule is domed with a mezzanine balcony to glimpse the first and second floors, archways open to the principal stairs, which in turn lead the visitor to the rear of the entry hall. This ultimate landing is the starting point for a clockwise route through the impressive collection of Renaissance paintings. The mezzanine landing has return flights for the second floor rooms. There are screened service cores linking basement to roof level which today contain passenger lifts for the disabled. The arrangement of one spacious area which opens on to another within the centre wing is masterly and enables the building interior to be revealed in perfect order. One is never lost as to which direction to pursue.

By contrast, the megalomania that overwhelms the Palais de Justice in Brussels dwarfs human scale and totally bewilders those who penetrate to the inner domed hall, a space that rises 40 m. It is said to be the largest edifice created in the nineteenth century, it could also be said to be one of the most complex layouts ever contrived (Figures 5.4a, b) – a veritable maze in the spirit of the dreaded Minotaur.

Contemporary examples are chosen from civic buildings where a central atrium or loggia has been seen as a unifying space to link the varying activities of the public.

The city hall of Åarhus constructed between 1937 and 1942 is the winning entry by Arne Jacobsen and Eric Møller in a competition. Many features simulate the fine detailing of Gunnar Aspland, in particular the glass enclosed lifts and the delicate scale of balustrading. The formal sequence follows a generous foyer placed below the Council Chamber, the connecting formal steps (reminiscent of Gothenburg in Figure 5.14) are sited on the main axis with balconies leading to the first floor suite (Figure 5.5a). A curving

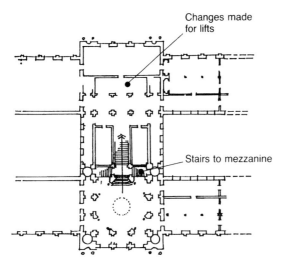

Figure 5.3 Kunsthistorisches Museum, Vienna, 1872–81. (Gottfried Semper)
a Outline plan at ground floor

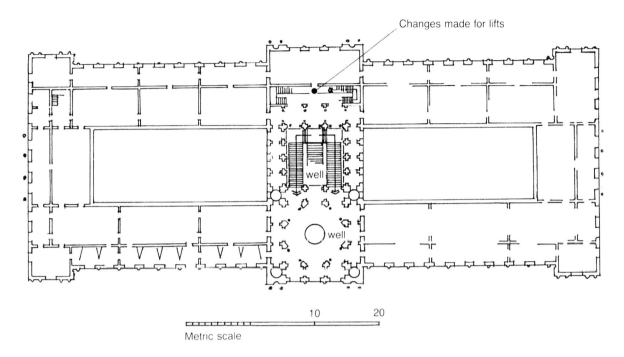

Figure 5.3b Outline plan first floor Kunsthistorisches Museum, Vienna

descending flight serves the cloakrooms and lavatories arranged out of sight in the basement (Figure 5.5*b*). The galleried atrium, which might unkindly be described as a modern day penitentiary, is serviced by separate stairs and glazed lifts housed within a visible core at the end of the hallway (Figure 5.5*c*). The contrast

Figure 5.3c View from vestibule

in finish and geometry between the richness of the entry and the simplicity of the office area is sufficient to mark the different zones. The use of balcony and suspended forms to dominate the foyer makes a perfect foil to the spaces served (Figure 5.5*d*).

Oslo City Hall, which is almost contemporary, was largely built in wartime and signified an emblem of resistance to Nazi occupation. The over-elaboration has to be seen as Norwegian self-expression in the face of adversity. The great inner hall is built on the scale of a Renaissance courtyard with a direct flight of gargantuan steps leading to the balconied upper floor, which is designed for an anti-clockwise movement through the public rooms (Figure 5.6).

Gottfried Böhm has a more elusive approach – the strata of his buildings are often unveiled layer by layer. The entrance

104 Stairs, Steps and Ramps

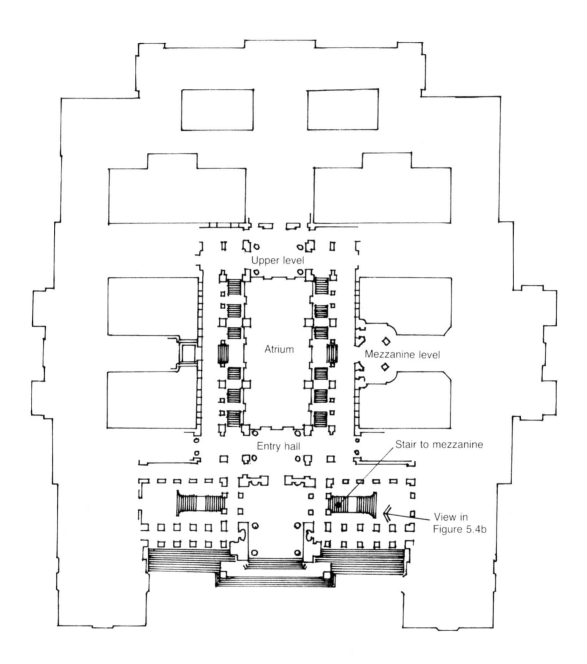

Figure 5.4 Megalomania writ large: Palais de Justice, Brussels, 1866–83. (Polaert)
a Plan of central staircase core

Figure 5.4b Main stair (from Fletcher, B., A History of Architecture, *Batsford, 1945)*

to the town hall of Rheinberg is through a generous open court, with a glazed conservatory sheltering double return stairs placed as a screen to the primary elements. Secondary accommodation to the wings either side have open escape stairs, which complete the external composition. (Figures 5.7*a, b*). The design is partly a reproduction of the Cortile to the Palazzo Municipio Genoa in modern dress. The duplication of stairways and entrances could be criticized but this small community building for Rheinberg does not have the security problems of the inner city. The formal and informal entries are placed one behind the other to suit the various municipal requirements.

Figure 5.5 Aarhus City Hall, 1937–42. (Arne Jacobsen and Eric Moller) a Contrast in forms

106 Stairs, Steps and Ramps

Figure 5.5b Curving stairs serving basement, Aarhus City Hall

Figure 5.5c Glass enclosed lifts and office stairs

By contrast Boston City Hall in the USA has a highly centralized approach designed for a greater level of security. There is guarded access from the basement and check points in the public hallways before entrance is permitted to the various departments and to official offices. The general arrangement above ground level has public walkways with ramps or stairs from City Hall Square and a sloping, brick paved, pedestrian area which links the site to Quincy Market and the waterside area. The node point is the landing space below the central light well, from this vantage point stairs and a ramp climb to the main entry sheltered by the bridge block of the upper storeys of Boston City Hall, the various elements being grouped in the four wings of the upper floors. The public circulation space allocated is well in excess of commercial offices but in this case the landings and vestibule are used to celebrate

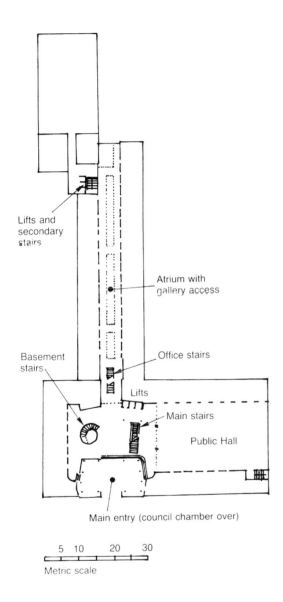

Figure 5.5d Layout plan, Aarhus City Hall

Figure 5.6 Oslo City Hall, 1937–50. (Anstein Arneberg and Magnus Poulsson) View of gargantuan steps within inner hall

the city, the space being dedicated to exhibitions, promotions and meetings. It is a form of city square under a canopy, complete with theatrical spaces, upper streets and balconies inspired by an Italian hill town (Figures 5.8*a*, *b*, *c*, *d*).

The theatrical quality of central stairs have already been assessed in terms of Garnier's Opera House and Dixon's proposals at Covent Garden (referring back to Figure 2.4 and 2.6). The stair hall and landings at TV/AM are conceived by Terry Farrell not simply as the central circulation space to the offices and studio space but as a set piece for filming television (Figures 5.9*a*, *b*). The movement left or right from the vestibule to the first floor is dramatized by landings or by groups of platform steps to gather people together. The range of levels permits variation in camera position whilst lighting effects can transform any location into a stage set. Lifts exist for general access from a lobby approach to assist sound proofing.

108 Stairs, Steps and Ramps

Figure 5.7 Rheinberg Town Hall (Gottfried Bohn)
a Internal view

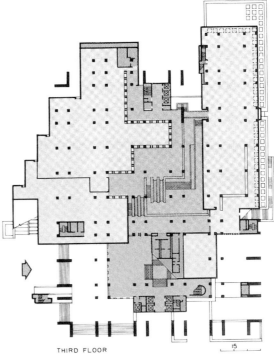

Figure 5.8 Boston City Hall, USA, 1962–69.
(Kallmann, McKinnell and Knowles)
a Key plan at lower level

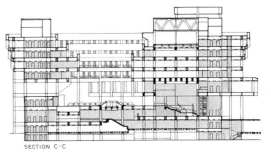

Figure 5.8b Key section

Figure 5.7b External view

Figure 5.8c View of entry vestibule at base of

Figure 5.8d Street level with theatre space, Boston City Hall

The unwinding of a central well stair can provide an excellent exploratory route in a small composition. The theme of a box within a box is one that occurs in the Rietveld–Schroder House (refer back to Figure 3.5) In the art gallery at Amersfoort the boxes are open spaces one within the other, the outer area forms the exhibition space while the stair turns the visitor to the first floor gallery with a circulation that terminates at a small cafe. The central well stair is the crucial element that allows the whole space at both floor levels to be fully exploited (Figures 5.10a, b). This theme of unwinding space is further developed

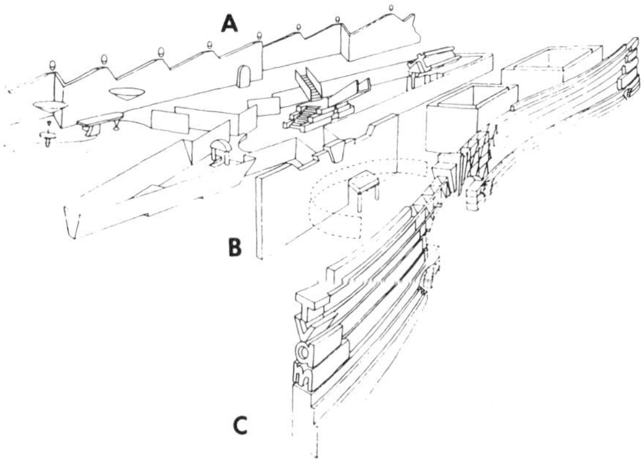

Figure 5.9 TV/AM offices and Studios, Camden Town, London, 1983. (Terry Farrell)
a Sketch of stair design

Figure 5.9b General view of landings and stairs

110 Stairs, Steps and Ramps

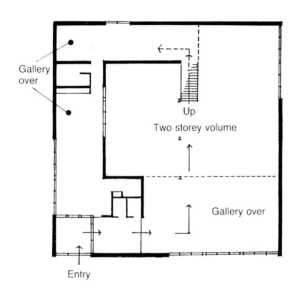

Figure 5.10 Exhibition Hall and Gallery 'De Zonnehof', Amersfoort compared with Van Gogh Museum, 1959. (Gerrit Rietveld)
a Key plan

Figure 5.10c Interior space surveyed by principal stairs at the Van Gogh Museum, Amsterdam, 1973

by Rietveld in the Van Gogh Museum in Amsterdam where the main stairs are placed so that the whole interior space can be perceived storey by storey (Figure 5.10c).

The pivotal role of staircases within Lasdun compositions has already been

Figure 5.10b General view of open well and turned stairs, 'De Zonnehof'

mentioned in connection with the National Theatre. The foyer space is discussed in Chapter 2, Section 5. It is worth studying the superb siting of stairs which permit the visitor to encapture the dramatic volumes that embrace the two principal theatres. The structural grid is placed at 45 degrees, thus allowing the stair shafts and vistas to be turned at sympathetic angles to the entries of the auditoria (Figure 5.11a). The diagonal axis is also the crux of the entry to the European Investment Bank, Luxembourg where the four wings of the construction are composed around a sequence of volumes which step diagonally down the hillside. The arrangement of angled stairs and chamfered planes and steps reflects the geometry of the diagrid coffers of the floors above (Figures 5.11b, c, d). Finally there is the unfolding geometry of the formal four-turn stairs within the central hall of the Royal College of Physicians, Regents Park (Figures 5.11e, f). To quote

Stairs, Steps and Ramps 111

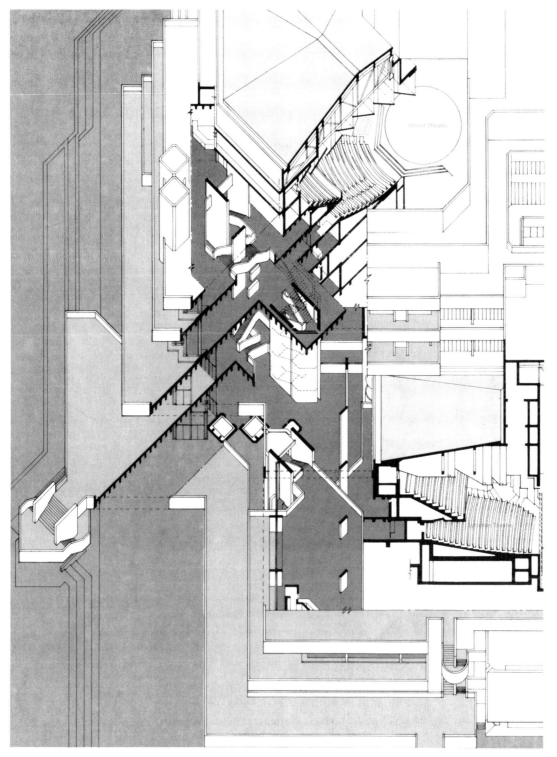

Figure 5.11 Lasdun's pivotal stairs
a Isometric of National Theatre Foyer Stairs, 1975

112 Stairs, Steps and Ramps

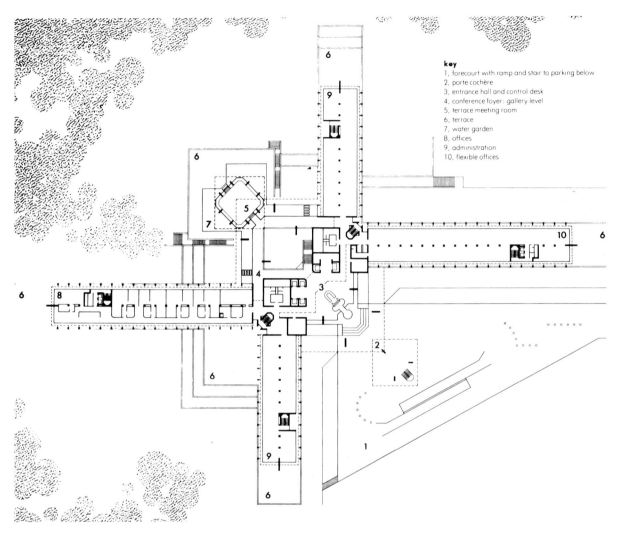

Figure 5.11b Diagonal axis stairs. European Investment Bank, 1981. Upper level

Lasdun, 'the language and theme of the formality is the essence of classicism'. Compare with the grand approach in the Palazzo Grimani (Figure 5.17).

The central theme can also be realized as a staircase spine ascending through the core of the plan. The Arthur M. Sackler Museum at Havard has a central light well slotted into the length of the block with direct flights rising towards the light (Figures 5.12a, b) the lighting is gained via a fenestrated wall towards the office area. The tubular handrail is lit on the underside to give dramatic illumination at night. Other stairs have been illustrated which cater for the human inclination to move towards the light or to the better lit spaces. Stirling and Wilford utilize this concept with the entry to the Clore Gallery, London, where stairs to the principal space rise within a roof-lit well (Figure 5.13).

Böhm has enlarged upon the device of stair screens at Zublin-Haus by providing

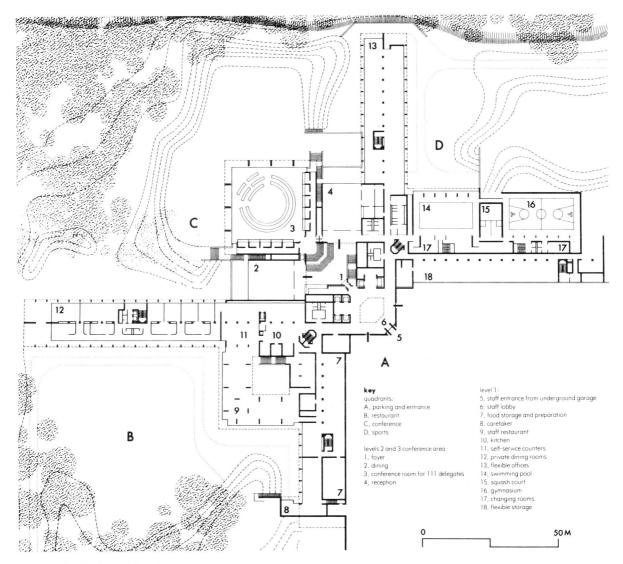

Figure 5.11c Lower level

such features as independent episodes within the vast atrium space. Fire stairs and service cores are discreetly placed within the enclosing wings of buildings with the lift shafts and public stairs brought into spatial prominence (Figures 5.14*a*, *b*). The IBM headquarters at Belfont Lakes featured in the case studies is a contemporary version of the same concept (see Case study 14.13).

5.3 Periphery stairs

Public stairs located to the edge of plans gives greater flexibility in disposing of elements within the total layout. In other words the axes can be left, right or centre without having to incur the penalties of symmetry. Some of the more dubious classical plans (Figure 5.15) suffer duplica-

114 Stairs, Steps and Ramps

Figure 5.11d *View of lower and upper stairs European Investment Bank*

tion of stairs where one would suffice. The examples chosen have been placed according to building form and are not in precise historic order.

First, the 'cortile' theme, with the Basilica at Vicenzia contrasted with the Palazzo Grimiani and the Courthouse at Gothenburg. Second, the periphery stair as an entry is seen in several instances in public buildings in London: Burlington House, the British Museum, the National and Tate Galleries. Examples of prostyle layouts are compared: the Maison Carrée, Nîmes and the Neue Pinakothek, Munich. Finally, Corbusian layouts are discussed and the influence on the Royal Festival Hall.

5.3.1 The inspiration of the 'cortile'

Allusions have already been made in Chapter 2 (Figure 2.11) to the fact that Palladio was not always perfect in the disposal of stairs. In fact the approach varied, at the worst one finds stairs relegated to

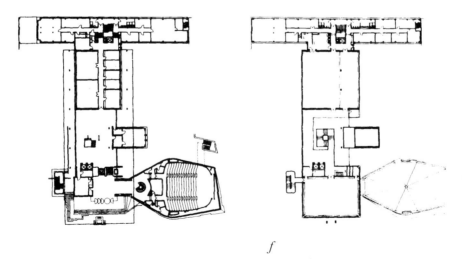

e   f

Figure 5.11 *Royal College of Physicians, Regents Park (1960) e Ground floor plan. (From* A Language and a Theme, *RIBA Publications 1976) f First floor plan. (From* A Language and a Theme, *RIBA Publications 1976)*

Stairs, Steps and Ramps 115

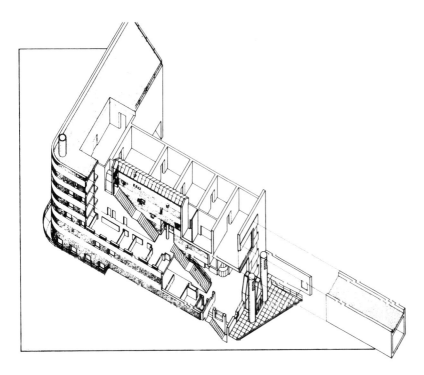

Figure 5.12 Arthur M. Sackler Gallery Harvard, USA 1984. (James Stirling, Michael Wilford and Associates)
a Diagram

Figure 5.12b Stair and light well

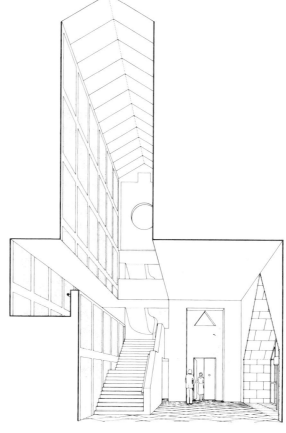

Figure 5.13 Entry to Clore Gallery, London, 1986. (James Stirling, Michael Wilford and Associates)

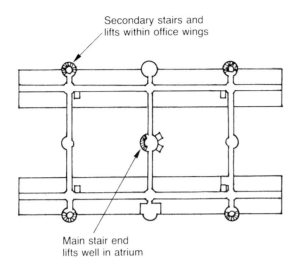

Figure 5.14 Zublin-Haus. (Gottfried Bohm) a Layout plan

Figure 5.14b Free standing screen of lifts and stairs in central space

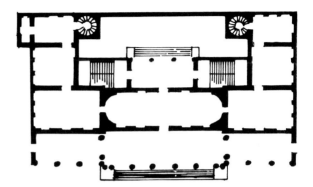

Figure 5.15 Needless duplication of stairs, Palazzo Chiericati (Palladio)

turret steps tucked away in corners (Villa Rotonda and Villa Thiene), on the other hand they seldom occupy a sequential space but are more likely to be adjuncts and placed in pairs where one major flight would be adequate, (Villa Godi and Palazzo Iseppo Porto). To be fair, the two storey loggias and stairs that encompass and transform the Palazzo della Ragione are a masterpiece in civic planning.

The outer circulation allows entries from any direction. It is another example of the 'master' space at the upper level surrounded by the 'servant' elements, this time a Council Chamber with loggias and direct flights of steps from the piazza (Figure 5.16). The steps are of medieval construction and cleverly incorporated within the enlarged composition.

The original idea may well have been taken from the 'cortile' relationship to loggias and stairs in the Venetian palaces. A common detail is the lack of rectangular form to the building periphery, the loggias to the Basilica at Vicenzia are an attempt to regularize the medieval geometry. In Venice the trapezoidal boundary is commonplace with regularity brought to bear by means of the 'cortile' and loggias with a stairway tucked away to one side (Figure

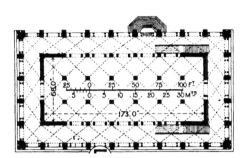

Figure 5.16 Palazzo della Ragione (known as the 'Basilica') Vicenza, 1549–1617. (Palladio) Ground floor layout with twin stairs and loggias. (from Fletcher, B., A History of Architecture, Batsford, 1945)

Figure 5.18 Courthouse extension, Gothenburg, 1934-37. (Gunnar Asplund) a Ground floor layout

5.17). The awkward shapes do not permit overall symmetry, the edgewise siting of the main stairs permitting considerable freedom in layout.

Asplunds extension to the Courthouse in Gothenburg is a modern solution to the same problem. The new wing is placed at the edge of the older courtyard building (Figures 5.18*a, b, c*). The entry is turned through a right angle, with the prospect of the new staircase rising behind a glazed screen. The axis is turned again towards the ceremonial route to the first floor courts and to the glass enclosed lift. The principal rooms are laid out to three sides of the stair hall on both floors, there is a separate stair at the end of the upper floor leading to offices for the court officials. The precedence of the first floor is announced by the single monumental flight within a two storey volume and signified by the greater ceiling height. The lighting levels to the staircase are also enhanced from the upper hall by clerestory lighting to balance the full height glazing to the courtyard. The single flight appears to lightly bridge the 12 600 mm span although it is cleverly suspended by steel tubular sections from the floor beams overhead. The gentle proportion of tread to riser (360 mm to 110 mm) and the gracious sweep of the handrailing without breaks in alignment produces one of the most perfect stairs to look at and enjoy in use. The siting of the free standing lift enclosure within the vestibule addressed the need of the handicapped in a gracious manner, a rare achievement in public buildings (See Figure 12.4*b*).

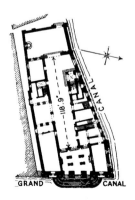

Figure 5.17 Palazzo Grimani, Venice, 1549. (Sanmicheli) Layout plan of the piano nobile. (From Fletcher, B., A History of Architecture, Batsford, 1945)

118 Stairs, Steps and Ramps

Figure 5.18b General view of formal stairs (see also Figure 12.4b)

Figure 5.18c Detail of secondary stairs

5.3.2 Periphery stairs as entries

Staircases at the extreme edge of a layout or arranged as a prominence are familiar in antiquity. The Propylaea Athens demonstrates the former while the Maison Carrée, Nîmes illustrates the latter – both have already been discussed in Chapter 2 (Figures 2.4 and 2.5). The continuing classical traditions finds these themes developed in significant forms. First the placing of entry stairs, often complicated by double or treble return flights on the contra axis of the composition (Figure 5.19*a, b, c, d*), sketches outline plans taken from various London public buildings of this pattern – Burlington House, the British Museum, the National and Tate Galleries. In all cases the functional requirement is to connect street level with the principal floor raised 4 to

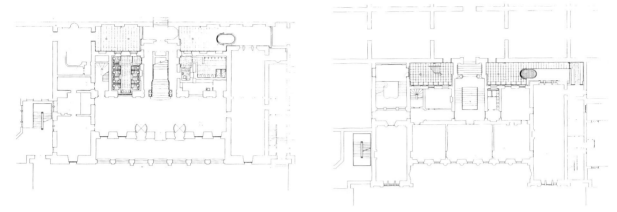

Figure 5.19 Outline plans of periphery stairs in the classical tradition around London a Burlington House, original work 1663–68, remodelling to main stair, 1816. (Samuel Ware) New Sackler Gallery stairs, 1990. (Sir Norman Foster)

6 m. It is significant that each of these public entrances have been left unaltered despite remodelling to the rest of the building. The original stair arrangements had sufficient subdivisions flight by flight or choice of direction to serve multiple accommodation. Ultimately, however, other edgewise entrance stairs have had to be introduced to serve the extended plan on the opposite side of the block at Burlington House and the British Museum. The same intrusions have occurred further along the main façade with both the National and Tate Galleries. On balance periphery stairs permit considerable variation to develop with the general plan and with the phasing of construction. In the case of the National Gallery, the grand entrance served perfectly well for over 140 years as the original site was infilled gallery by gallery. By comparison a central space stair on the lines of the Kunsthistorisches Museum, Vienna (Figure 5.3) occupies a fifth of the whole plan. This would be unsupportable if only a quarter of the gallery space had been constructed in the first phase, an impoverished situation that occurred with the initial work at the National and Tate Galleries.

5.3.3 Prostyle layouts

The entry to the Maison Carrée is the derivative for the second pattern of classicism in staircases, namely the entry axis arranged parallel to the length of the building. The Neue Pinakothek, Munich (Figure 5.20) takes the concept to the utmost degree. The length of the gallery is dictated by the fixed collection of exhibits. The entrance is via impressive steps housed within the gable end, the internal sequence is formed by a long succession of galleries placed in a loop. No extension is possible, the growth of the museum area at Munich is dependent upon a series of separate pavillions. More familiar applications are the eighteenth-century churches of London and New England where the 'prostyle' plan has a portico and steps to grace the gable elevation (Figure 5.21). In modernist terms, a latter day architect K.J. Schatner has substituted the lift for the staircase in ordering the symbolic portico to the School of

120 Stairs, Steps and Ramps

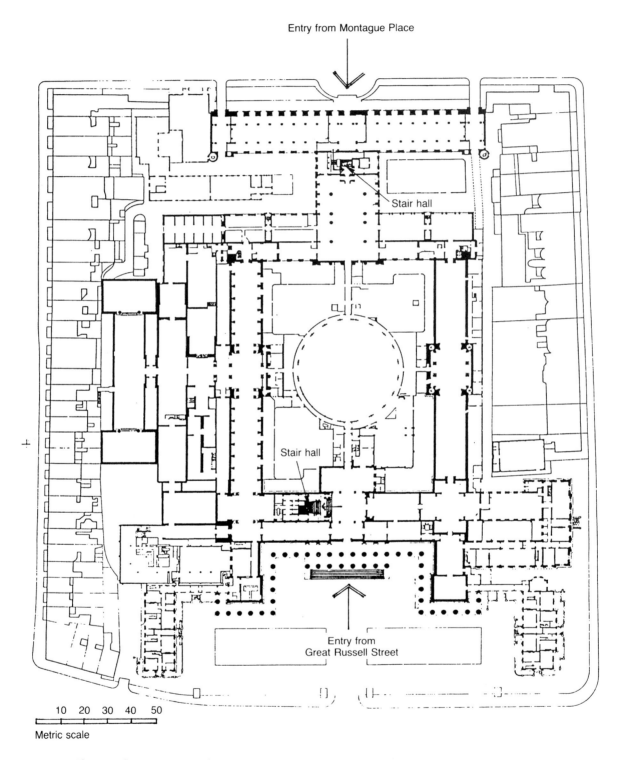

Figure 5.19b British Museum south entrance, 1842–47. (Sir Robert Smirke) North block 1904–14. (Sir John Burnet)

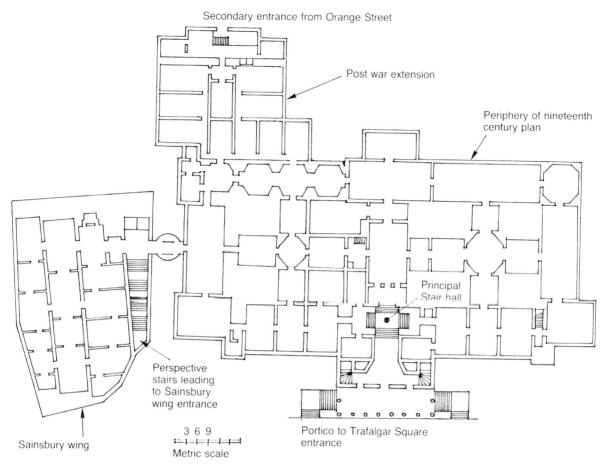

Figure 5.19c National Gallery, first phase, principal stairs and adjacent frontage 1832–8. (William Wilkins)

Journalists at Eichstatt (Figure 5.22). Here the design of the entry door glazing bears a cryptic reference to the lift shaft beyond. The circular shaft and slot for the lift appear as a façade pattern. Once in the building, the shaft is presented as a column with a slot like opening to the lift cage. The stair is totally suppressed and placed in a secondary position.

5.3.4 Corbusian layouts

It is necessary to turn to Le Corbusier for a distinctive phase of modernism in the placing of vertical circulation. The League of Nations Competition (1927), The Centrosoyus Building, Moscow (1933) and the Palace of the Soviets Competition 1931 (Figures 5.23*a, b, c*) are three crucial corbusian designs which illustrate the development of open planning in public spaces. A common feature all these civic or public buildings is the open foyer from which stairs and lift shafts rise to principal elements above. In the case of the Palace of the Soviets the form of the main auditorium is visible as a floor print within the ceiling structure. There are large open wells where staircase forms can be seen as ascending sculpture set between the '*piloti*'. The placement of the stairs orders the movement

122 Stairs, Steps and Ramps

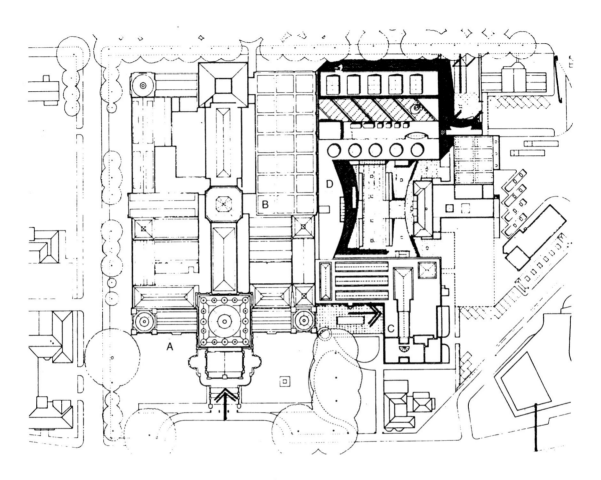

A Original phase 1890 (followed by 1907 and 1935)
B Postwar gallery
C Clore Gallery
D Final phases to be built

Figure 5.19d Tate Gallery, London, various phases 1907–1990

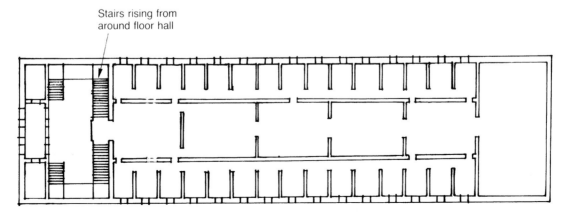

Figure 5.20 Neue Pinakothedk, Munich, 1846–53. (August von Voit) Outline plan

Figure 5.21 St Martins in the Fields, 1722–26. (James Gibbs) Portico and steps

pattern and places the vertical circulation to the edge of the spaces served. The relationship to the upper floors is nodal with stairs and lifts becoming the hinge points between the office accommodation.

X and Y related forms have already been outlined in Chapter 4. These forms can make powerful compositions when dramatic stairhalls are placed visibly at the interphase of the main wings. The UNESCO Centre in Paris and Nestlé Headquarters at Vevey are both supreme demonstrations of this theme (Figures 5.24*a*, *b*, *c*). The Nestlé composition has greater visual appeal with the ground floor lifted on elaborate *piloti* with a clear view to Lake Geneva beyond. The transparency given to the stair halls ensures equally attractive views outwards from the open circular stairs and related hallways.

A popular building with fine public spaces is the Royal Festival Hall, where the Corbusian image is achieved with remarkable clarity[3] (Figures 5.25*a*, *b*, *c*, *d*). The staircases are naturally lit and placed beyond the edge of the auditorium. The stair location has enabled the open plan ground floor to be changed around, in fact, the entrances have switched from side to side. The upward sequence through the various levels is a fine promenade. It is also an experience where the 'sense of place' within the whole circulation is well assured. The design of the stairs is enhanced by good lighting, by day- and night-time, the simple palette of first class materials has also contributed to the detailing which appears dateless 40 years later.

5.3.5 Conclusion

A philosophic debate on whether stairs should have central, nodal or peripheral status cannot be concluded with a handful of examples. Further case studies are given at the end of this volume with the argument developed in differing directions. There can be little doubt that the three dimensional creativity that characterizes architecture from building is

124 Stairs, Steps and Ramps

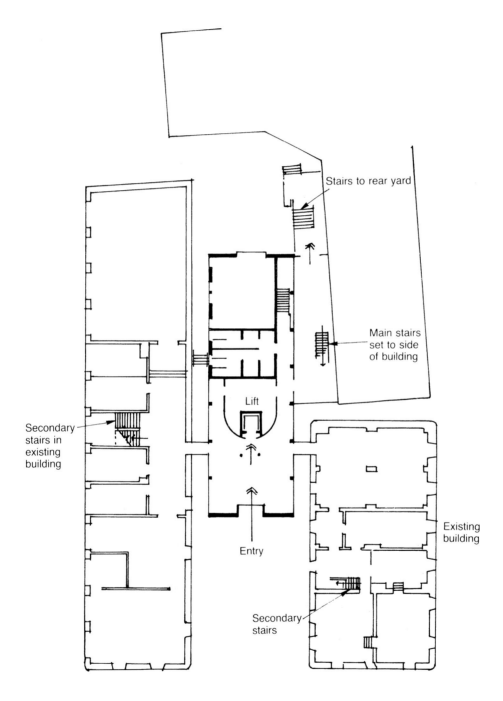

Figure 5.22 School for Journalists, Eichstatt, 1988. (K.J.Schatner) Lift replacing stairs as symbol of entry. Stairway in secondary position

Stairs, Steps and Ramps   125

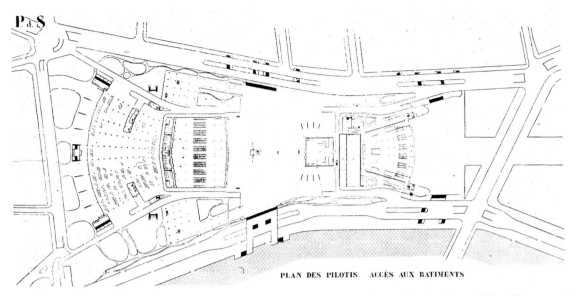

Figure 5.23 Corbusian layouts and their derivatives (from Le Corbusier Architect of the Century, Arts council, *1987)*
a League of Nations Competition, 1927

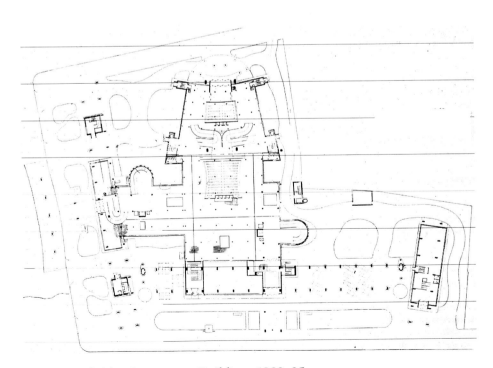

Figure 5.23b The Centrosoyus Building, 1929–35

Figure 5.23c Palace of Soviets, 1931

signified by the design skill displayed when handling staircases. The concluding illustration (Figure 5.26) is one of the finest examples, the ceremonial stairs designed by Jože Plečnik for the Hradcany Castle, Prague.

Figure 5.24 Nestlé HQ, Vevey (Jean Tschumi) a Interior

Stairs, Steps and Ramps   127

Figure 5.24b Elevation

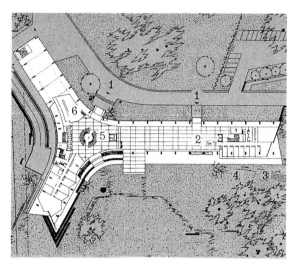

Figure 5.24c Ground floor plan. (From The New Nestlé international headquaters in Vevey*)*

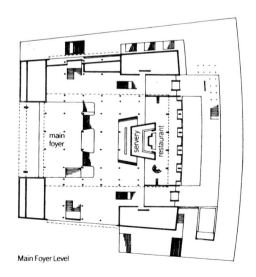

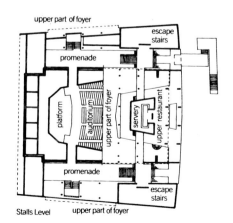

*Figure 5.25 Royal Festival Hall, London, 1951. (Sir Leslie Martin and Sir Robert Matthew)
a Key plans (stalls and foyer)*

128 Stairs, Steps and Ramps

Figure 5.25b Stair in relation to reception, Royal Festival Hall

Figure 5.25c Stairs in relation to foyer

References

[1] Both Parliament buildings need to be looked at in detail, the most revealing texts can be found in the *Architectural Review* for February, 1950, which describes the rebuilt House of Commons and compares this with the West German Parliament buildings, completed in Bonn at the same time. More recently the *Architectural Review* (March, 1993) has featured the Bonn parliament complex built for a united Germany.

[2] Further examples of Lasdun's pivotal stairs can be found illustrated as follows: Royal College of Physicians (*Architectural Review*, April, 1965); National Theatre (*Architectural Review*, January, 1977); European Investment Bank (*Architectural Review*, November. 1981); Milton Gate, City of London. (*Architectural Review*, June, 1991); *A Language and a Theme*, (RIBA Publications Ltd. 1976).

[3] There was a recent re-appraisal in *Architects Journal*, 9 October, 1991.

Figure 5.26 Ceremonial stairs Hradcany Castle, 1920–22. (Jože Plečnik)

6 External stairs

6.1 Garden architecture

Some aspects of designing external stairs come within the realm of garden architecture. It is not possible to condense that vast topic, the reader is therefore left to search out their favoured source, Oriental or Islamic, English, French, Italian or Spanish or perhaps that blend of vernacular that purtains to European landscape design today.[1]

There are common forms of external steps despite the stylistic range already referred to. A primary role concerns steps that relate to buildings, a secondary aspect are those which serve the functional needs for access into or through the landscape or townscape. Both forms have developed a decorative role in garden architecture, like steps placed around a pool or in a kerb, or plinths to mark differing levels, and the decorative aspect can be enhanced by the use of materials, such as grass versus paving.

External stairs carry symbolic undertones with perhaps a greater poignancy due to the connection with nature. There are many temples of the Buddhist faith that are constructed in the form of steps to heaven. One Chinese example is the sacred road up the mountain of Tai Shan in Shantung. Here a flight of ancient granite steps (refer back to Figure 2.1g) with lesser shrines and pauses leads the pilgrim through a long ascent which can be seen in holistic terms as the ascent of life itself. Southern India and Sri Lanka have similar mountain shrines (Figure 6.1a). The most famous is Adam's Park, south-east of Colombo, it rises to a height of 2 240 m with a conical summit that terminates with an oblong platform. Within the platform, there is a hollow, resembling the imprint of a human foot though measuring 1 600 mm × 750 mm. The footprint is ascribed to Adam, to Gautama, and to Siva and held in veneration by Muslims, Buddhists and Hindus alike. Heavy chains on the south-west face, are said to have been left there by Alexander the Great, and mentioned by Marco Polo as indicating the original ascent.

The reverence inspired by a temple mount can also pertain to more modest themes, as witnessed by the Kennedy Memorial, Runnymede. It is a place where a gentle stepped path through the woods concludes on the open hillside with a simple block of limestone for the

Figure 6.1 Symbolic undertones a Hindu shrines on hillside ascent Chamundi, Mysore, India. (Courtesy of Timothy Blanc)

dedication and a foot space set in the sward (Figure 6.14) The indentation of the final steps within the slope is reminiscent of Ancient Egypt.

A differing response occurs with steps that lead down into sunken chambers or that disappear below water. The psychology of descent into shade or below ground or below water are total opposites to the optimism in climbing Tai Shan. Some significant examples are the cistern temples in India where the architectural ensemble is a preparation for the ceremonial cleansing at the lowest level. The pattern is developed from the holy 'ghats' seen at Benares, where four miles of river terraces line the Ganges for Hindu pilgims (Figure 6.1c). The cisterns and water steps which feature in many temple complexes are designed for large crowds of worshippers within a sacred enclosure (Figure 6.1b).

The theme of descent and purification was captured superbly by the Architect to the 'Memorial de la Déportation' at the eastern end of the Ile de la Cité (Figure 6.1d). Here one descends from a green garden down stone ramps into a sunken shaded courtyard of masonry and iron with glimpsed views of the Seine forever breaking past the promontory, turn round and there is the inner sanctuary, a peaceful softly lit hall of remembrance. The return route is back into the light and the hustle and bustle of the market behind Notre Dame.

Figure 6.1b Holy Ghats at Benares, with stepped terraces leading down into the water

c
d

*Figure 6.1c Water steps and tank Temple at Chidambaram, India. (Courtesy of Timothy Blanc)
d Memorial de la Déportation, Paris, 1962, sculpture by Desserprit. (Georges-Henri Pingusson)*

6.2 Historic precedent

The outdoor room is probably the most pertinent description that holds true for the vernacular English Garden.[2] The precedent stretches back in European terms to the Italian Renaissance and in turn to the gardens created in Southern Spain at Granada and Seville. In essence, the outdoor space is compartmented into 'rooms and passages' laid out across sloping land with stairs to encapture movement from level to level.

Stepped paths are matched with stepped parapets in Granada (Figure 6.2a) in contrast to pebble ramps enlivened by fountains bordered by walls having waterfall copings (Figure 6.2b). Spanish gardeners accompanied the Borgias when they transferred themselves to Italy, little wonder that the stepped water gardens in Tuscany, Rome and its hinterland or Venetia have the qualities of Andalusia. The quintessence must surely exist in the terraced and multi-staired water paradise at the rear of the Villa d'Este (Figure 6.3a). This garden, which serves as an outdoor room, is the size of Trafalgar Square and criss-crossed with three vistas in each direction, with ramps and staircases that tumble down 20 m from the palace to the furthest terrace. Each step and turn

a   *b*

Figure 6.2 Moorish inspiration in the Generalife Gardens Granada, fifteenth century; restored in 1920s
a Stepped paths and parapets
b Pebble ramps, enlivened by fountains bordered by walls having waterfall copings

is water embellished (Figures 6.3*b*, *c*, *d*), each fountain considered as a musical accompaniment, each movement down the multi-turn stairs compared say, to a minuet (Figure 6.3*d*). A comparison of Italian garden stairs to dancing steps is not far-fetched. Plans of the Spanish Steps, Rome (Figures 6.4*a*, *b*) convey in their arabesques and curves the notion of a musical cadence. The concept of turning, stopping, advancing and retreat is the key to the elaboration achieved within this Italian tradition. The symbol of dance and lovers' meetings is captured by the sublime music of Le Nozze di

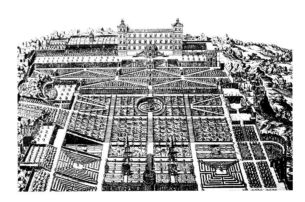

Figure 6.3 Villa d'Este, Tivoli, 1550s. (Designer Pierre Ligorio)
a Key plan

Figure 6.3b Return stairs enclosing a fountain, Villa d'Este

Figaro where the penultimate scene occurs within an Italian garden.

A similar cadence occurs at Villa Garzoni and Villa Medici (Fiesole) where designers have translated the outdoor space into a veritable palace of garden rooms with some of the grandest external stairs ever constructed[3] (Figures 6.5a, b). The key visual element in Italian gardens is the role played by stairs as scenery to frame the views, where the diagonal or curving balustrade patterns provide the essential clues as to direction and geometrical composition. The landscape proportion of riser to tread is often as gentle as 100 mm × 400 mm as occurs in the Spanish Steps (refer back to Figure 6.4c). Landscape architects are taught to use differing proportions to interior designers, usually twice the rise plus the tread to equal 700 mm instead of 600 mm.

Another detail of Italianate design is the way the spatial experience is played upon or pinched at stair locations (Figures 6.6a, b, c, d, e, f). Perspective effect can be increased by tapering the flight or by reducing the balustrade size in distant views. Expectations can be raised by using curved platform steps at gateways, either sunken or raised flights, a favourite device of Edwin Lutyens. Dull, repetitive parallel treads can be enhanced by slightly curving the nosing edge (in plan), a detail referred to by Sir George Sitwell in his treatise 'On the Making of Gardens'. (Figure 6.6e) illustrates a curved nosing line taken from the terrace steps at Renishaw. The use of a footspace at

Stairs, Steps and Ramps   135

c

d

*Figure 6.3c Stepped ramp, waterfall coping and gargoyle
d General approach from palace*

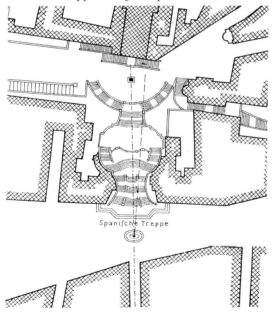

*Figure 6.4 Spanish Steps, Rome, 1721–23.
(Designer Francesco de Sanctis) a Key plan.
(From Schuster, F., Treppen Hoffman Verlag,
1949)*

the base and head of garden steps will prevent wear and tear in a lawn setting, the geometry of this foot space also assisting with orientation.

6.3 Steps outside buildings

Returning to mundane matters the National Regulations in Britain call for ground floors to be at least 150 mm above the natural exterior level, hence a step is needed at all entries (Figure 6.7a) or else a ramped pavement where disabled access is mandatory (Figure 6.7b). The 1992 changes in the National Building Regulations call for textured

136 Stairs, Steps and Ramps

Figure 6.4b General view, Spanish Steps

pavings at ramps to identify such features for the blind and partially sighted. Both elements can be combined in a single design (Figure 6.7c). The dimensions of the basic form (Figure 6.7a) should be in scale with the door, the depth should be a comfortable footspace. A more generous approach would be of person girth, desirable in domestic and public entrances. Gibberds steel 'prefabs' were perfectly thought out for courting couples, *circa* 1945, with a 1 800 mm × 900 mm step covered by a porch complete with a balustrade to lean against! (Figure 6.7d).

Providing a flight of steps by external doors needs to take into account safety of egress, space for door swings and a walking distance before the first riser, say a distance of 1 800 mm minimum (Figure 6.8a). Balustrade protection should be given to both sides, or handrails provided[4], where the platform

Figure 6.4c Detail of gentle riser to tread proportion

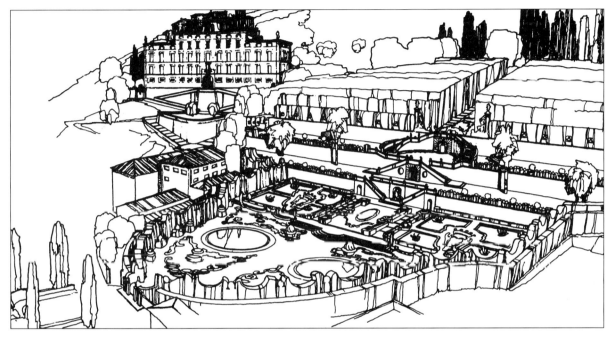

Figure 6.5 A veritable palace of garden rooms a Villa Garzoni Collodi, 1652

Figure 6.5b Villa Medici, Fiesole, 1458–61. (Michelozzo Michelozzi) (source of both figures from Italian Gardens of the Renaissance by J.C.Shepherd & G.A.Jellicoe. Princeton Architectural Press (1986)

138 Stairs, Steps and Ramps

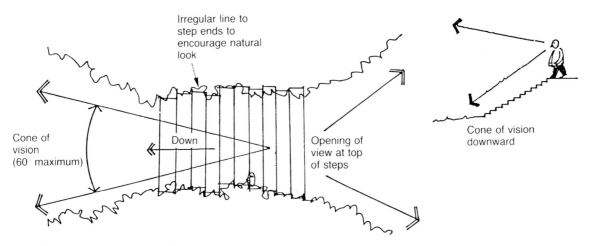

Figure 6.6 Italianate detail
a Tapered layout for perspective effect

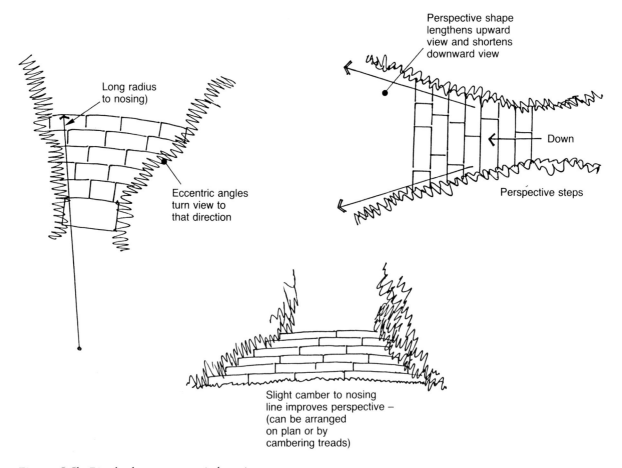

Figure 6.6b Pinched space at stair locations

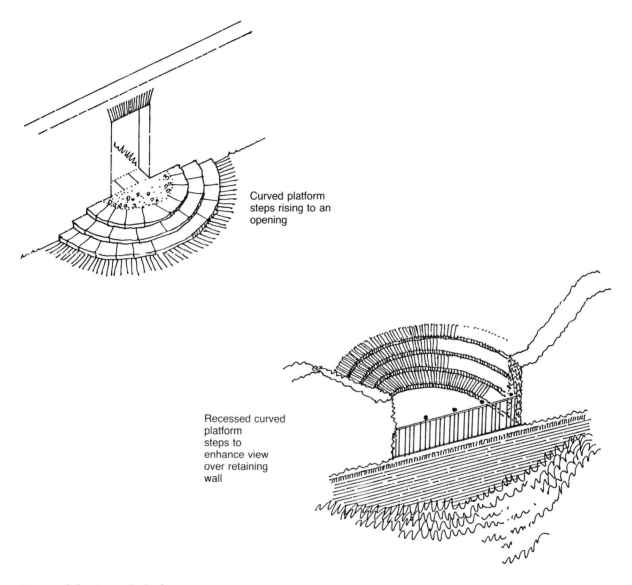

Figure 6.6c Curved platform steps

category of stairs is employed. Balustrading constructed with tubular steel can incorporate electrical conduit and lighting systems placed within the handrail profile. In many European countries ramp blocks are built into the flight (Figure 6.8b) but in Britain ramped circulation has to follow a defined slope of given width. The imposing civic entrance to the Law Courts in Vancouver is a very successful application of these principles on a monumental scale, the complex levels combine a sitting out area with diagonal ramps and stairs for ceremonial purposes (Figures 6.8c, d). The concept relates to platform steps employed in pyramid form with both external and internal corners. The main advantage is

Figure 6.6d Differing heights of balustrade to give greater distance, Powys Castle, date unknown

Figure 6.6e Curved line of tread to improve visual effect, Renishaw, 1900s. (Designer Sir George Sitwell)

the prominence that such forms give to entries (Figures 6.9a, b). The simpler application has origins that rest with the stylobates of Greek temples (refer back to Figures 2.5 and 2.6). Direct flights of stairs are more practical and easier to adapt for safety with guard rails and perimeter balustrading (Figure 6.9c) and with ramps accommodated alongside. Direct flights can also be devised to handle vast crowds of people, as in the layout of sports stadia. In these circumstances the steps are divided into passage-ways of 1800 mm width separated by engineered guard rails, with groups of stairs limited to 16 risers between landings,[4] failure can produce a catastrophe. In Britain, such stairs have to break their alignment at every third landing. Similar provisos exist when designing the approaches to pedestrian bridges or underpasses, the customary layout having a choice of stairs and ramps.

Reverting to simple steps placed by building entries, there are severe constructional restraints that affect the design solution. By illustration, the plinth or block of steps can be made as a ground slab independent of the building mass (Figure 6.10a), or else as a cantilever extension of the ground floor slab or adjacent wall (Figure 6.10b). The other concepts involve independent framing in steel or timber (Figures 6.10c, d). By comparison the Greek stylobate depended upon a rock foundation, which presented no movement between elements of a structure that were heavily or lightly loaded (refer back to Figure 2.5).

The final picture (Figure 6.11) demonstrates the fallacy of ignoring ground conditions. Frank Lloyd Wright in his prairie designs followed Mayan concepts with

Stairs, Steps and Ramps 141

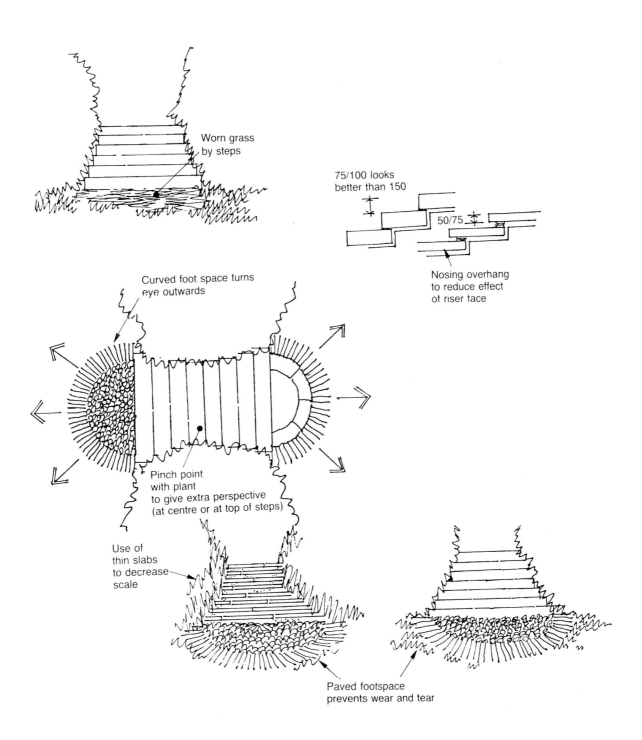

Figure 6.6f Footspace in relation to garden steps

142 Stairs, Steps and Ramps

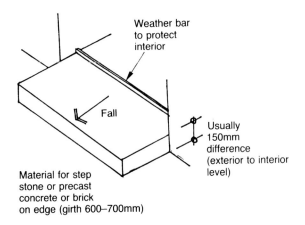

Figure 6.7c Combined pattern: new entry steps at RIBA HQ, London, 1980s. (J. Carey)

*Figure 6.7 Simple steps outside buildings
a Basic form of door step*

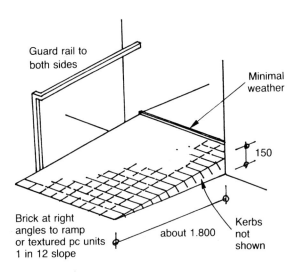

Figure 6.7b Basic ramped entry American codes call for a 50 mm height kerb at the sides to restrain wheelchairs, 100 mm kerbs in the UK

building and garden features (like plinths or massive steps) constructed in one plane. A hundred years later and the Wrightean plinths are revealed to be wildly out of line due to settlement trouble. A point of detail which the Mayans and the Greeks who built temples and temple steps of solid rock did not suffer.

Figure 6.7d Raised step, porch and railing, BSIF Prefabricated Houses, circa 1945. (Sir Frederick Gibberd)

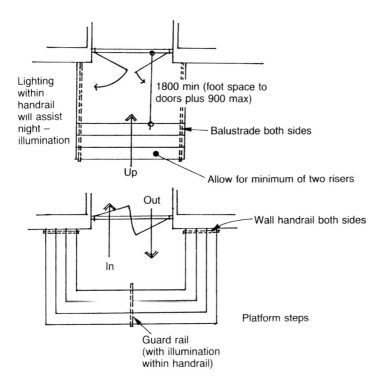

Figure 6.8 Direct flight steps
a Basic dimensions

6.4 Steps in landscape

The most extensive ramped and stepped construction ever made is the roadway which surmounts the Great Wall of China, the only human construction said to be visible from the moon. The design embraces a defensive wall, backed by a 4.5 m paved surface, either ramped or set out as stepped ramps in the steeper sections. The materials employed are largely brickwork. The undulating lengths involve all forms of inclined surfaces, ramps, zig-zag climbs, curved and wide spaced steps, a source of ideas for every other form of hilly path in the landscape (Figures 6.12*a, b, c*).

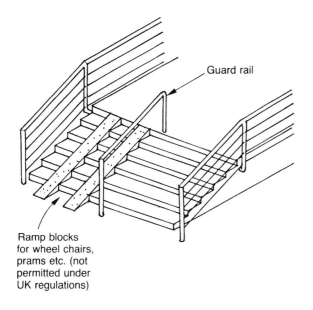

Figure 6.8b Ramp blocks

144 Stairs, Steps and Ramps

c

d

Figure 6.8c and d General and detail views: ceremonial stairs and ramps at the Courts Complex, Vancouver, 1972–79. (Arthur Erickson)

Stairs, Steps and Ramps 145

Figure 6.9 Platform steps and large scale stairs a Semi circular form, Mausoleum of the Aga Khan, Aswan, Egypt

Figure 6.9b Splayed form at entry to Federal Offices, Seattle, USA, 1974. (Landscape architect Richard Haag)

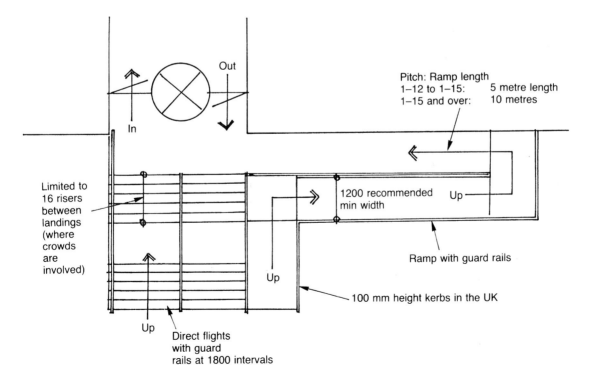

Figure 6.9c Use of guard rails and balustrades for direct stairs and ramps at entries. (See Figure 11.2a for table regarding strengths and heights)

146 Stairs, Steps and Ramps

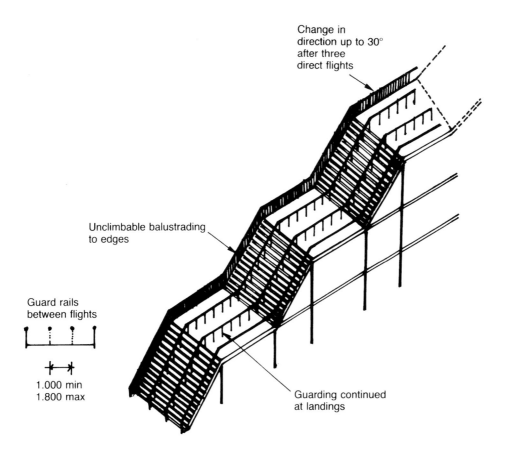

Figure 6.9d Large scale external stairs in sports stadia

Figure 6.10 Constructional restraints on design
a Independent slab steps   b Cantilever slab stairs

Figure 6.10c Stair platform in steel

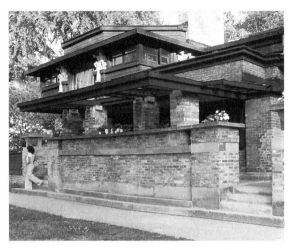

Figure 6.11 Settlement problems: Wrightean plinth cracked at junction between house and garden features. (Courtesy of Curt Teich & Co. Inc.)

Figure 6.10d Stair platform in timber

The conversion of private gardens into a public domain can cause problems – not least the lack of guard rails as seen at Lindisfarne Castle, now owned by the National Trust (Figure 6.12*d*). At present this detail is covered by a warning note and adequate insurance. Matters were just as problematical after Lutyens' improvements had been completed in 1908. The following is a quotation from a letter to Lady Emily concerning a visit by the Prince of Wales (later George V). The Prince 'was terribly alarmed at the gangways up and wanted a wall built. I told him we had pulled one down and that if he really thought it unsafe we would put nets out. He thought that very funny'.

Monumental steps in the landscape are usually associated with Baroque grandeur – Caserta, Chatsworth, San Souci and Versailles are illustrated as typical (Figures 6.13*a*, *b*, *c*, *d*, *e*, *f*). The main vista at Caserta is almost a mile in length, constructed as a canal with waterfalls and water steps decorated by fountains and sculpture (Figures 6.13*a*, *b*). It is certainly inspired by Grillet's cascade at Chatsworth, a fraction of the scale deployed at Caserta but still faithfully capturing the delight of water steps as an umbilical thread in landscape (Figure 6.13*c*). The cascade of steps below Sans Souci were devised to spread crowds through the terraces that step down the hillside. The terraces had continuous

148 Stairs, Steps and Ramps

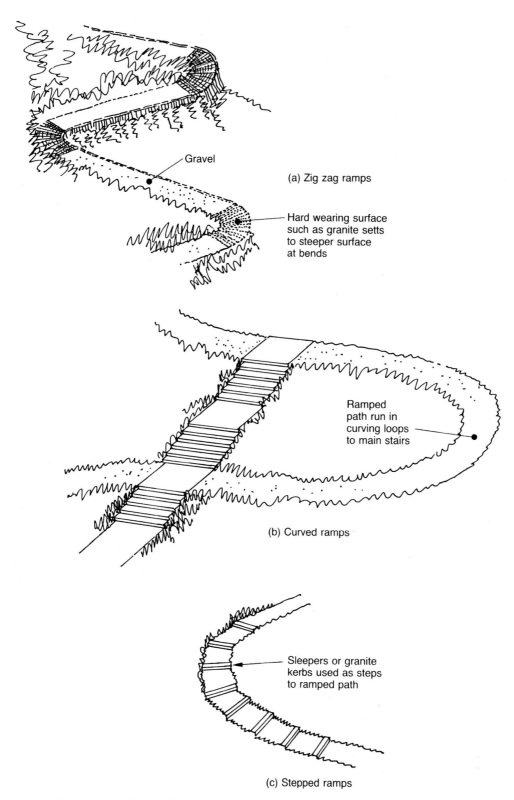

Figure 6.12 Forms of stepped paths

Figure 6.12d Stepped ramp, Lindisfarne Castle, 1911. (Sir Edwin Lutyens)

glazed frames for fruit growing, old photographs reveal the contrast of the multi-paned conservatories rising as a glass ziggurat between the fan shaped steps (Figure 6.13*d*). The layout below Sans Souci was of Dutch and French origin and one has to look at Versailles to discover the grandest sequence of all.

The Sun-King's domain portrayed by the park at Versailles has within it the symoblic elements of the state, the farm, forest, lake, river and the city, represented by the enfolding wings of the royal palace. The terraces provide a vast dais as to a throne, the edges become castle walls broken by vast stairs that open the view to the horizon (Figure 6.13*e*). The detail that links each parterre within the 'dais' are kerb profiles that match the treads

Figure 6.13 Monumental steps in landscape
a General view, water steps and waterfalls, Palazzo Reale, Caserta, 1752. (Designer Vanvitelli)

150 Stairs, Steps and Ramps

Figure 6.13b Detail of water steps, Caserta

and risers to the terrace steps and in turn to the plinth mould of the Palace of Versailles (Figure 6.13*f*). A superb monumental detail of steps in landscape architecture.

Garden steps play a far more modest role, to show the way or to give emphasis to a particular vantage point. The Kennedy Memorial at Runnymede demonstrates the most eloquent expression of steps and surfaces in a simple landscape setting. The approach is through woodland with the pathway picked out in granite setts, the steeper

Figure 6.13c Water steps, Chatsworth, Derbyshire, 1694. (Grillet (pupil of Le Notre))

parts having ramped steps. The open meadow beyond the wood has mounded setts that rise to the climax of the memorial. The formal approach route is made via sunken stone treads of ample proportion, these relating to the terrace paving on which is mounted the plinth block and memorial stone (Figures 6.14a, b).

6.5 Steps in townscape

An essential primer has been presented in Chapter 2 concerning the Capitol, Rome (Figures 2.15a, b, c), a further reminder is given with the Spanish steps (Figure 6.4a). The accidental delight of steps in the context of the street-scape is the other pleasure (Figure 6.15a) and one that is captured by the architect Sir Clough Williams-Ellis in the romantic Italianate village of Portmerion (Figures 6.15b, c). The car is virtually forbidden, which means that the pedestrian theme is paramount. The details are carefully attended to and much of the material used is salvaged stone, even slate 'blanks' from WC seats of the one and two holer pattern. The scale and texture

152 Stairs, Steps and Ramps

Figure 6.13d Fan shaped steps below San Souci, Potsdam, 1745–47. (Designer C.W. von Knobelsdorff)

Figure 6.13e Great staircase, Versailles, 1661–81. (Designer Le Notre)

Figure 6.13f Detail of plinths and steps

Figure 6.14 Kennedy Memorial, Runnymede, 1965. (Sir Geoffrey Jellicoe)
a Ramped steps

has a Cullenesque quality but predates Gorden Cullen's beguiling sketches by many years (Figure 6.15*d*).

Other sources of inspiration are the civic spaces created by Lawrence Halprin as part of a pedestrian network within the cities of Portland and Seattle. The Portland plan comprises a link between the University area and the civic buildings and the edge of 'downtown'. A plaza has been created at either end of the walkway route, where steps play a major role in the urban landscape (Figures 6.16*a*, *b*). Halprin has woven together the movement patterns around Lovejoy Plaza with space for water and a gazebo with ramps placed around the edge of the space to enable

154 Stairs, Steps and Ramps

Figure 6.14b Formal steps and memorial

Figure 6.15 Steps in townscape
a Steps in Prague, Czechoslovakia   b General view, Portmerion, 1920s. (Sir Clough Williams-Ellis)

Stairs, Steps and Ramps 155

Figure 6.15c Detail of steps at Portmerion

full participation by the community. The lower Plaza by the Civic Auditorium is a visual affair with water steps and fountains for the fearless (Figure 6.16c).

Figure 6.15d Sketch by Gordon Cullen from 'Townscape'

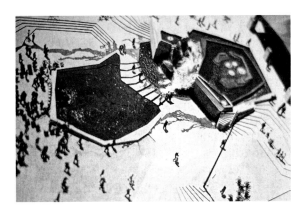

Figure 6.16 Pedestrian Plazas, Portland, Oregon, 1967–68. (Lawrence Halprin)
a Layout plan of Lovejoy Plaza

Figure 6.16b Detail of steps, Portland

Figure 6.16c Water steps in South Auditorium Plaza and waterfall

References

[1] *Garden History & Garden Design.* Further Reading. There is no shortage of books and guides to stimulate ideas, the following selection are those which are best documented concerning garden steps.
Oriental. *The Chinese Garden* by Maggie Keswick (Academy Editions, 1978).
Spanish. *Spanish Gardens* by Marquesa de Casa Valdes. (Antique Collectors Club, 1987).
English. *Houses & Gardens by E.L. Lutyens* by Lawrence Weaver, (Country Life, 1913).
French. *The French Garden 1500–1800* by W.H. Adams. (New York, 1979).
American. *Lawrence Halprin*, published as Volume No.4 by Process 1978.

General Reference. *Landscape of Man* by Geoffrey & Susan Jellicoe. (Thames & Hudson 1975).
Modern Garden by Peter Sherherd, (The Architectural Press, 1953).

[2] *The Edwardian Garden* by David Ottewill. (Yale University Press, 1989).

[3] Refer to descriptions of Italian gardens in the following
Italian Gardens of the Renaissance by J.C. Shepherd and G.A. Jellicoe. (Princeton Architectural Press, 1986).
Italian Gardens by Georgina Masson (Thames and Hudson Ltd., 1987).

[4] For strength of guard rails refer to Figure 11.2*a* in Chapter 11.

7 Detailed construction: timber

The intention in this chapter is to look at constructional ideas rather than the 'hands on' approach inherited from the nineteenth century where every method of setting out raking treads or wreathing to handrails are portrayed.[1] There are many excellent carpenters' manuals that provide this backup, today, reinforced by computer-aided design (CAD) packages used by designers and manufacturers. Figures 7.1 a–e are intended as a reminder of the tactile quality associated with timber stairs, an explanation is given in Section 7.7 for the selection made.

Figure 7.1b The most perfect example of cased stairs made by Ottoman craftsmen for the pulpit to the Ibn Tulun Mosque, Cairo, 1296.

Figure 7.1a Ladder steps made from planks, circa 12th century, Norway

7.1 Framing

(Figures 7.2a, b, c, d, e, f, g)

The most primitive stairs relied on foot holds chopped into a log, the next stage was the ladder form still used in adventure playgrounds (Figures 7.2a, b).

160 Stairs, Steps and Ramps

c

d

Figure 7.1c Elegant wooden flight reinforced by a wrought iron flitch within the strings, late 18th century, Stamford Assembly Rooms
d Spiral stairs called the Miraculous Stairs built without nails, 1878, Our Lady of Light Chapel, Sante Fe, New Mexico

Figure 7.1e The ultimate in timber elaboration. the main stairs at the Gamble House, Pasadena (balustrade modelled on the 'lifting cloud' motif of Japan), 1908. (Henry and Charles Greene)

Figure 7.2 Framing timber stairs
a Log pole with foot holds chopped out, circa 12th century, Norway

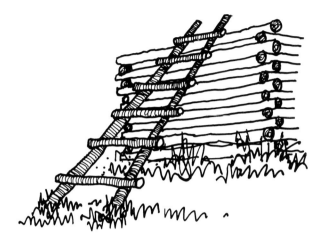

Figure 7.2b Log ladder circa 20th century, childrens playground

Figure 7.2c Medieval form of log treads, used in conservation work at Singleton, Sussex

Robust treads can be cut from quartered logs and secured down to a carriage timber, a medieval detail that will interest conservationists (Figure 7.2c). Ships ladders reveal considerable refinement with members whittled down to give the most efficient use of timber. Such construction is the prototype of today's open tread stairs (Figure 7.2d). The stability relies on effectively connecting the outer strings in the absence of overlapping rungs or treads of the more primitive pattern. The essential elements can be summarized as follows:

- Uncut strings to maximize longitudinal strength.
- Solid plank or multi ply treads with adequate glue line to mortices.
- Steel cross ties with plate bolt ends to tie the strings together.
- Refinements such as part risers will assist with the load bearing quality of the treads. This will increase the glueline and help comply with the British Regulations concerning the maximum 100 mm gap.

The width to span proportions are dictated by engineering and the economics of timber supply. Open tread flights in ladder form are usually limited to a metre width and for runs up to 16 risers. The sizings given in Figure 7.2d are for domestic purposes, say 900 mm width × 14 risers.

Wider designs, up to 1 800 mm, will involve carriage pieces with cantilever treads or a combination with steel brackets to fully support long span treads, shown in Figures 7.2e. Cut string open stairs appear clumsy and present problems in effectively framing treads and strings, such construction is more appropriate to composite

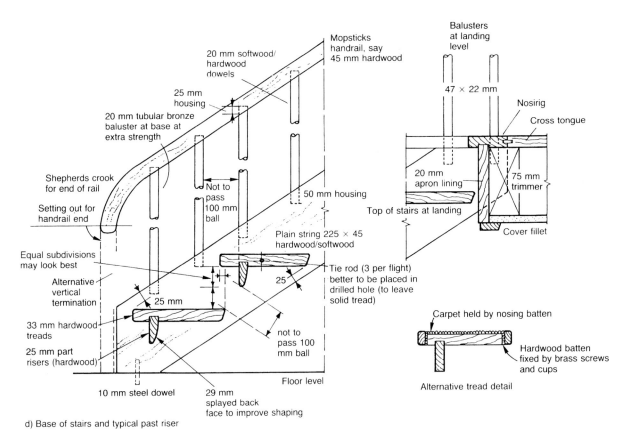

Figure 7.2d Details of open tread stairs

assemblies. Narrow ladderlike steps are permitted in the UK for access to a single room, refer to Chapter 3, Section 3.2.4 for details. Conventional framing relies on 'box construction' where a set of stairs is 'cased' together, hence the term staircase. Mass production and preferred dimensions have ensured that costs are well below custom made designs. Staircase kits from suppliers include a range of modules that come with all the ancillary detail – newels, balustrades, handrailing and trim for platform treads and return nosings where cut strings are preferred. It is also possible to directly assemble dog-leg and multiple turn flights from standard components. Figure 7.3 is a typical page from a manufacturer's catalogue.[2] Compare that to the superior quality of profile where an experienced architect is involved as with the replacement stairs in Figure 7.4b. The newel post has a framing role in conventional wood framed stairs, it provides a bearing plane for flights and handrailing as well as masking jumps in handrail alignment (refer forward to Figure 7.5e). In times past such work was entirely framed in oak enriched with carvings and statuary as at Hatfield House (Figure 7.4c), perhaps the finest Jacobean staircase in England. More comely versions exist in the Inns of Court and at the former offices of the Architectural Press in Queen Annes Gate (Figure 7.4a).

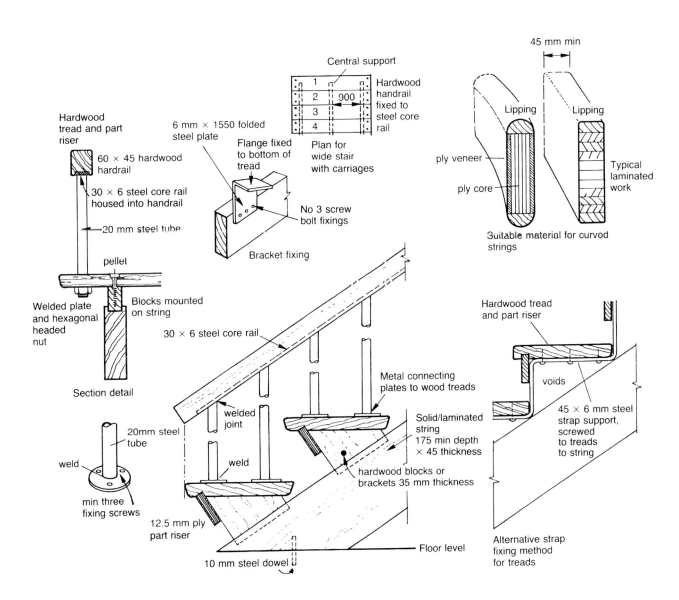

Figure 7.2e Open treads with carriage pieces showing details for blocks, brackets and straps. (See Figure 7.9c for carriage construction)

164　Stairs, Steps and Ramps

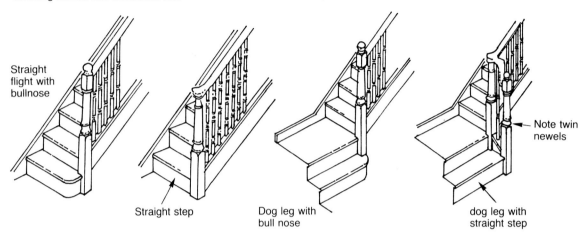

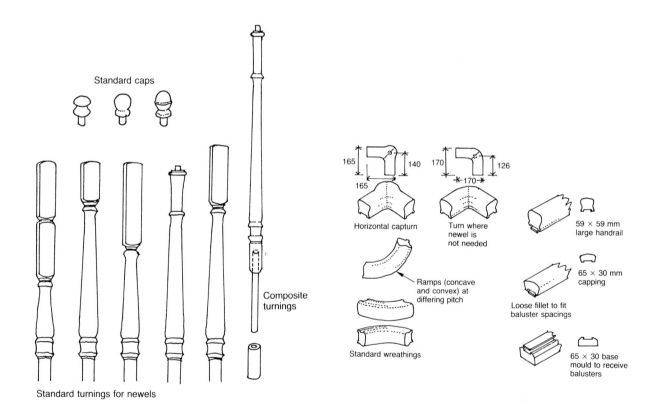

Figure 7.3 Typical page from manufacturers catalogue. (Redrawn by author). (Courtesy of Richard Burbidge & Son Ltd)

Figure 7.4a Stairs in Queen Annes Gate, London, 1704. (From Cruickshank, D, and Burton, N., Life in the Georgian City, *Viking, 1990)*

b

Figure 7.4b Replacement stairs after fire damage, Morton House, Highgate, London, 1990. (Julian Harrap)

7.2 Curved framing

The continuous line of curved stairs is more enticing than the interruptions caused by newels. Timber framing in the form of laminated strings assisted by a wrought iron flitch plate date back to the eighteenth century, (refer back to Figure 7.1c) though carpenters often propped their work with a slender iron column as extra security. Otto Salvisberg at Roche Chemicals, Welwyn Garden City put in a chromium plated tube to camouflage his belt and braces attitude to curved strings rising 3800 m (Figure 7.5a)

The construction resembles the open tread stair except that the strings are made from glued ply or blocks (like

c

Figure 7.4c Hatfield House stairs, 1620

laminboard) with face veneers and lippings. The integral strength comes from the tread acting with the inner and outer strings (Figure 7.5b). An alternative strategy is to use a central post, either solid or laminated, as a drum to carry the tapered end to each tread (Figure 7.5c). A more elegant solution found in Central Europe is to carve a curved newel into a hollow half cylinder. This shape fulfils the role of a newel in supporting treads, as well as the inner strings and handrail (Figure 7.5d) without destroying the line of the balustrade. (Compare Figure 7.5e for a traditional solution.) Carpenter's manuals from the eighteenth and nine-

Figure 7.5 Curved framing
a Curved flight, Roche Chemicals, Welwyn Garden City, 1938. (Otto Salvisberg). (From Schuster, F., Treppen, Hoffman Verlag, 1949)

teenth centuries are worth studying, Figure 7.5*f* revealing the geometry of oval stairs similar to the elegance in Figure 7.1*c*.

7.3 Non-traditional framing

Wooden treads can be suspended from the trimming timbers of the floor above, a feature that Walter Segal developed for the self builders at Lewisham. It was first demonstrated in his own house at Highgate in 1964 and is detailed in Case study 14.4. Another Segal innovation is the design of a wooden newel cage from the top to the bottom of a well; refer back to Figure 3.22*b* and to the connection with Ottoman architecture. Box construction with plywood is another option for self-building which enables winders to be constructed without relying on traditional style kits, the general arrangement is given in Figure 7.6. The plywood is fixed to the envelope walls behind a dog-leg stair whilst the boxed newel is set up for framing the winders, the remaining straight flights can be assembled from open treads or cased in with tread and riser as required.

168 Stairs, Steps and Ramps

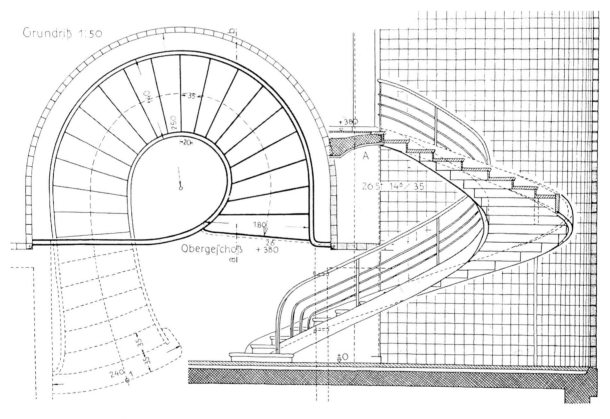

Figure 7.5b Detail of circular stair with central support. (From Schuster, F. Treppen, *Hoffman Verlag, 1949)*

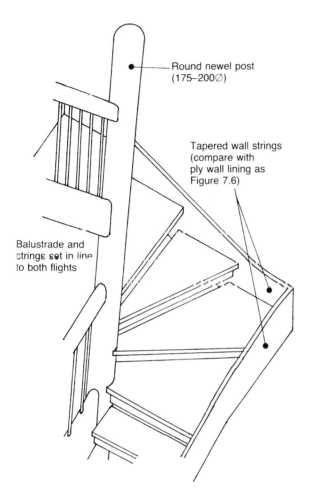

Figure 7.5c Detail of winders with post support

Figure 7.5d Curved newel to support strings and handrail, Wood House, Shipbourne, 1938. (Gropius and Fry, detailer Bronek Katz)

7.4 Tread detailing

The traditional 'cased in' stair has a nosing edge dictated by the housing joint of the riser (Figure 7.7a). A further complication occurs with cut strings where the nosing profile is reflected as a trimming feature to each exposed step (Figure 7.7b). Carpet finishing in such circumstances implies painted margins and the unsightly look of the selvedge behind the balusters. Straight strings avoid these complications (Figure 7.7c).

A complaint about open timber tread stairs is their noisiness in use and the difficulties that arise in coverings. Replaceable nosings will permit a range of finishings to be applied, the treads being considered as individual trays (refer back to Figure 7.2d).

7.5 Balustrade detailing

A debate on the rigmarole of British Building Regulations is set forth in

Figure 7.5e Traditional newel post

Chapter 11, not to mention the British Building Inspectors 100 mm spherical ball. Patterns based upon 100 mm intervals may pall and it is worth considering varied spacing to improve the scale or to vary the girth of the baluster. In practical terms it is an easy matter to construct a vertical rod balustrade in timber provided adequate purchase exists in the string, and landing nosing. (Refer back to Figure 7.2d. For tread ends in cut strings refer back to Figure 7.7a). Horizontal rail designs are simpler to frame up with newel posts (Figure 7.8a) and make a stout barrier if threaded through with tubular steel (Figure 7.8b). In the same vein, the inclusion of metal rod verticals within a timber balustrade will considerably stiffen the construction. Another age old device, from observing stairs in Georgian houses, is the discreet metal bracket that pins the underside of the handrail to a fixing on the apron lining (Figure 7.8c). Mesh and sheet panels usually require metal framing to give adequate strength where the choice rests with proprietary systems (Figure 7.8d).

7.6 Handrailing

The traditional kit stairs mentioned in Section 7.1 (Figure 7.3) have the attraction that standardized wreathings and terminal blocks are available, even though the shapes have the coarseness of Edwardiana. Equal economy can ensue if standard mopstick handrails are used without wreathing. Once turned work is needed then the services of specialists will be required to fabricate wreathed work at stair wells and for terminal blocks, etc. The handrailing connections are made by special bolts (Figure 7.8e); various attempts have been tried to market a universal timber handrailing that will solve all circular work (Figure 7.8f). Metal fabricators have come nearest to this ideal and would appear to hold the key to composite designs where timber stairs are combined with mesh or sheet balustrading.

The comfortable feel of timber to the touch should not be forgotten. Handrails are shaped to be grasped, the simple mopstick (45 mm diameter) or oval (64 mm × 40 mm) are immensely comfortable and much superior to the brutalist bulks favoured in the 1960s (Figure 7.8g).

Stairs, Steps and Ramps 171

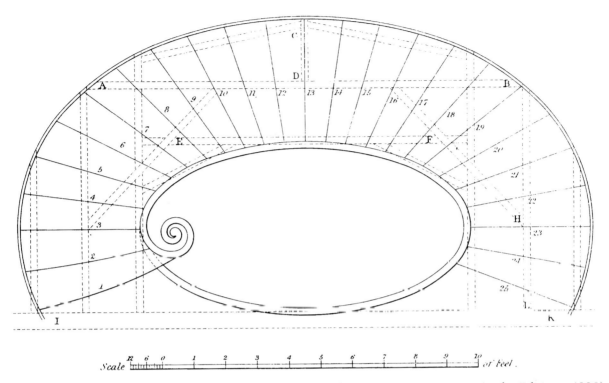

Figure 7.5f Georgian elliptical stairs. (From Newland, J., The Carpenter's Assistant, *Studio Editions, 1990)*

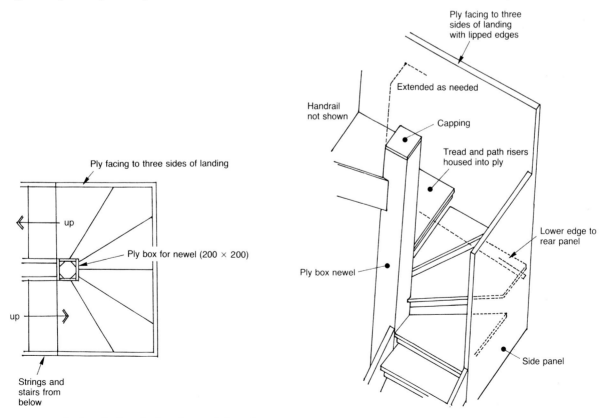

Figure 7.6 Sketch details for plywood framing

7.7 Trimming detail

Timber stairs should preferably be fitted after plastering in order to protect the woodwork and to provide a movement joint between the wall string and the finished surfaces (Figures 7.9a, b). The fixing of wooden flights is a matter of propping against trimmers with a housing joint, the base can easily be secured by dowel or coach bolt. A timber apron lining surmounted by a capping will provide a suitable edge trim to wooden floors. The advantage of a wide capping is the better fit that can be made for balustrade fixing, apart from the opportunity to match the profile of the treads at the well location (refer back to Figure 7.2d). Landings will need trimming to support wide treads and winders as Figure 7.9c.

7.8 Tactile quality

The tactile quality of timber can be sensed from the lead-in pictures. It is a matter of form and texture which gives as much pleasure to the maker as to the user. One of the most perfect examples is the work involved with the cased stairs to the 'mimbar' or pulpit from the Ibn Tulun Mosque, Cairo (Figure 7.1b). The timbers are Turkey Oak and Mediterranean Pine, the latter used for the single plank of carved balustrading. This feature is hinged downwards to improve vision lines when the prayers are led by the Imman.

The 'Miraculous Stairs' (in Figure 7.1d) were apparently built without nails by a travelling carpenter who came and went without being paid, another sample of devoted work. The nuns at the little

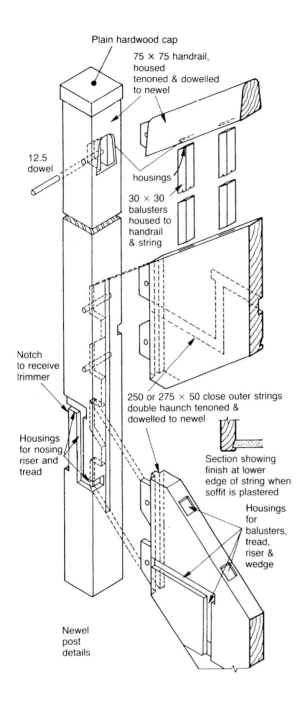

Figure 7.7 Principles of cased construction
a Traditional detailing

Complex shapes were developed by Aalto to provide anthropomorphic forms to fit the hand (Figure 7.8b) but the cost of wreathing is formidable.

Stairs, Steps and Ramps 173

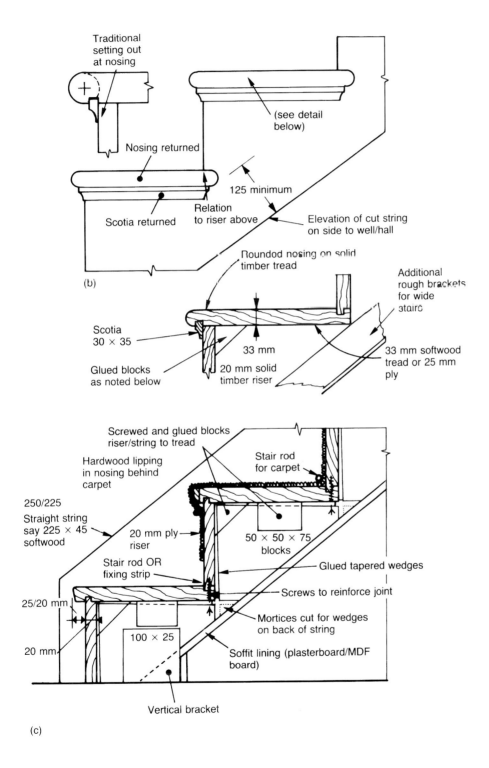

Figure 7.7b Cut tread nosing
c Nosings for straight string stairs

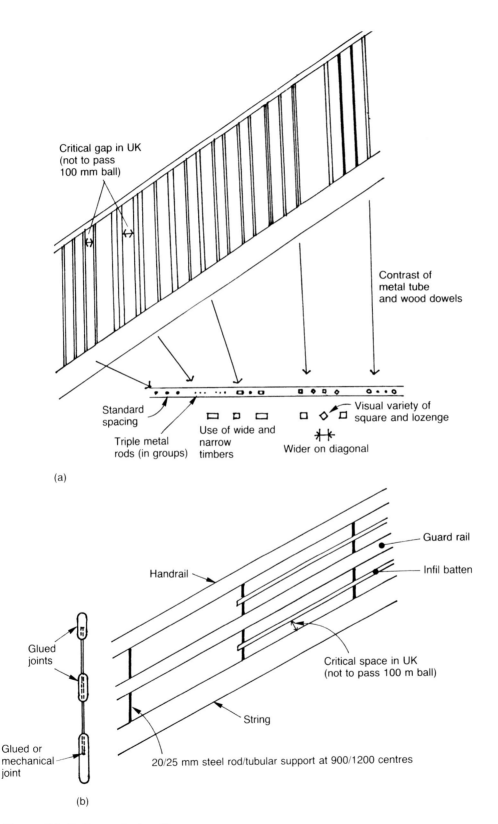

Figure 7.8 Ballustrade detailing
a Vertical and horizontal rail balustrade b Steel rod supports to rails

Stairs, Steps and Ramps 175

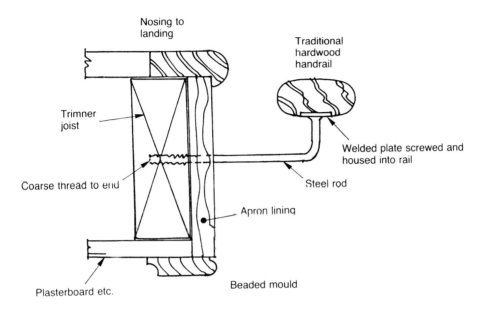

Figure 7.8c Discreet handrail bracket

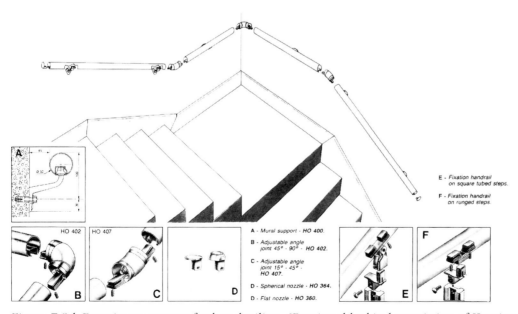

Figure 7.8d Proprietary systems for handrailing. (Reprinted by kind permission of Kensington Traders Ltd)

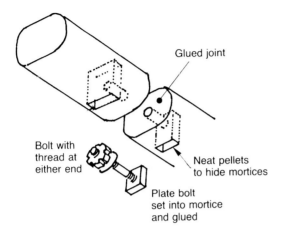

Figure 7.8e Handrail bolts

church called 'Our Lady of Light Chapel' Sante Fe still whisper under their breath that the itinerant stairmaker was called Joseph.

Miraculous or not, there is little doubt that the USA was provided with a great range of inspired work by European emigrés. The Gamble House, Pasedena (according to the guide book) owes the finishing work entirely to European skill. The aesthetic direction of the designers Greene and Greene may be Japan (Figure 7.1*e*) but the enveloping skill that assembled the handcarved stair encapsulates the joy of working in wood that is found in Alpine Chalets. Modern techniques with laminboard and ply can be equally inspiring as portrayed in a spiral stair designed by Luisa Parisi (Figure 7.10*a*), an example of lineal sculpture in space and equal to the invention of Victor Horta (Figure 7.10*b*). The geometry of propped and suspended construction has intrigued many designers. An interesting

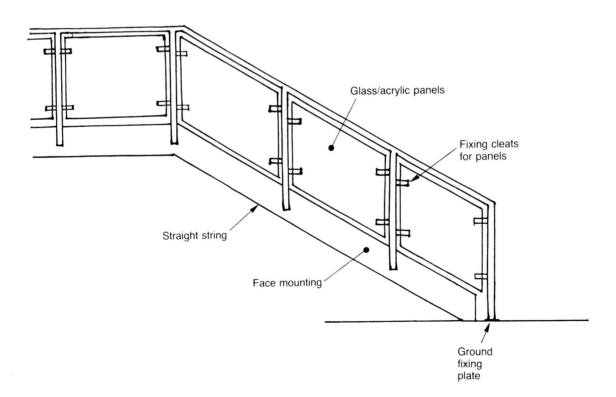

Figure 7.8f Universal balustrading: Hewi system using metal tubular sections and panels

Stairs, Steps and Ramps 177

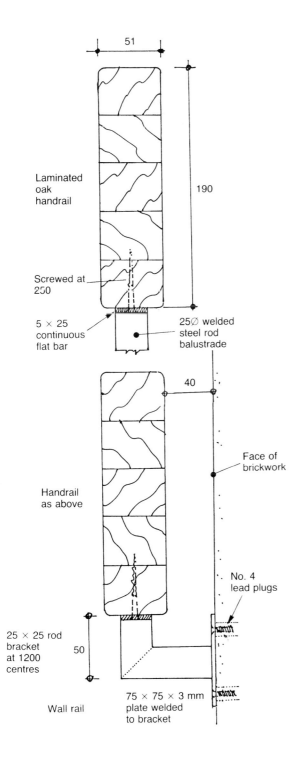

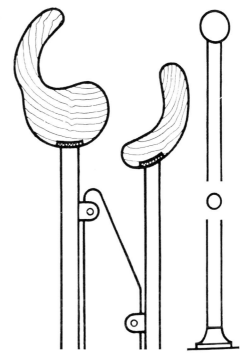

Figure 7.8h Aalto's anthropomorthic handrails. (From Schuster, F., Treppen, Hoffman Verlag, 1949)

concept is featured in the concluding illustration (Figure 7.11) which combines both features within one staircase. The traditional cased stair was removed in a cottage conversion and new framing made in the form of shelf construction. The lower most element is a book case with stair treads to the back, whilst the suspended upper flight is made from balustrade boards extended downward to support the last few treads.

Figure 7.8g Brutalist profile

178 Stairs, Steps and Ramps

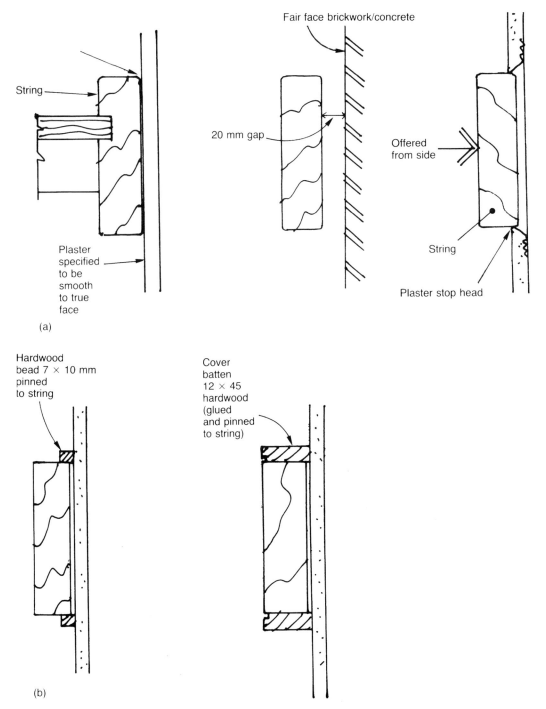

Figure 7.9 Trimming details
a String and plaster movement joint   b Covermoulds

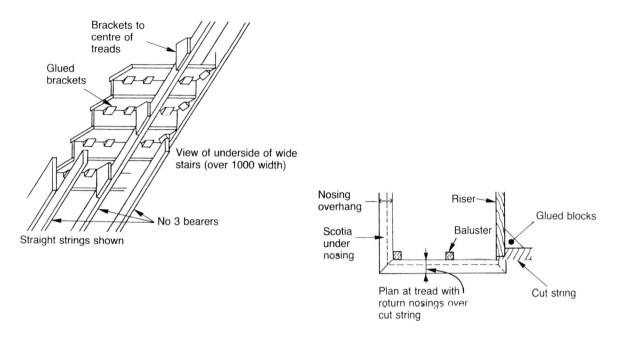
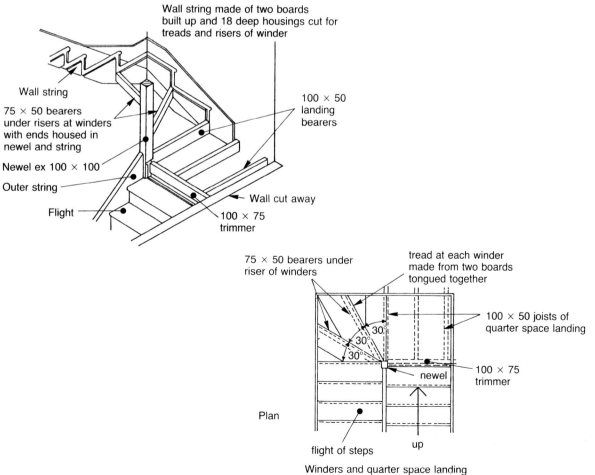

Figure 7.9c Trimming at landings and winders. (Redrawn from The Construction of Buildings *Vol 2, Barry, R. Blackwell Scientific Publications, 1992)*

180 Stairs, Steps and Ramps

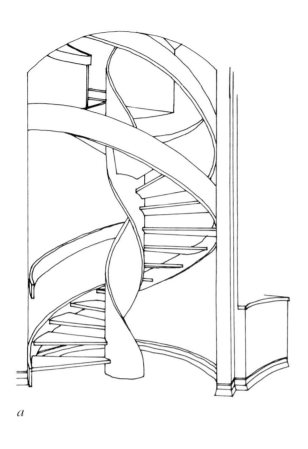

Figure 7.10 Lineal sculpture in space
a Plywood stairs by Luisa Parisi, 1973
b Victor Horta's handrail and newel post at his house in Brussels

Figure 7.11 Stair framing considered as furniture at Walmer, 1975. (John Bruckland)

References

[1] Two Carpenter's Manuals are *The Carpenter's Assistant* by James Newlands (in facsimile by Studio Editions, 1990)
The British Architect or the Builders Treasury of Staircases by Swan (1775)

[2] Domestic staircases are very much part of the 'bricolage' which governs house building both sides of the Atlantic. In the UK the leading components suppliers for stairs are Richard Burbidge and Son Ltd, Botrea Stairs (Saxondell Ltd) and John Carr Joinery Sales Ltd all of which produce almost identical catalogues.

8 Detailed construction: iron, steel and other metals

8.1 Ladders and ladder steps

Cast- and wrought-iron work in ladders and stairs have a tradition that stems from the Industrial Revolution (Figures 8.1*a, b*). Variations can still be seen in London 'areas' (Figure 8.2*a*); another source is the spider's web of steps that straddle the Eiffel Tower. The lattice girders were designed as ladders to enable the repainting to be carried out without cradles or scaffold. The only proviso made was the employment of unmarried men to save compensating the next of kin in case of a fatal accident.

Steel construction is also associated with maritime architecture and ships stairs. Ship shape forms are certainly the inspiration for Figures 8.1*c, d*.

Figure 8.1b Copper coated wrought-iron stairs, Stock Exchange Building, Chicago, 1893–94. (Adler and Sullivan)

Figure 8.1a Cast iron sectional spiral stairs, Palm House, Kew Gardens, London, 1844–48. (Designers Turner and Burton)

The addition of lightly framed steel stairs and handrailing can be less intrusive than masonry construction when making ruins accessible for tourists. At Rochester Castle, John Winter was asked to insert floors and add a roof to the open shell of the ruined tower. He was also asked to construct a staircase in the old position so that the public could enter the castle through the original doorway. It is not known how the original stairs or roof

c

d

Figure 8.1c Glass block and steel stairs, Waterloo Terminal Development, 1992. (Nicholas Grimshaw and Partners) (Courtesy of Jo Reid and John Peck)
d Trussed stairs: 'Legends' night club, London, 1987. (Eva Jiricna, Engineers Dewhurst MacFarlane)

were constructed, therefore, it was decided to make these frankly modern structures, but to keep them discreet so that the famous views of the castle were not affected. The staircase is a simple structure of steel bars and plate with oak treads. It is finely detailed however and makes a neat foil against the old stonework (Figure 8.1e).

8.2 Fire escape stairs

Fire escape stairs are a more familiar pattern and where iron and steel are still employed in fabrication. The traditional construction relies on steel strings (formerly plate) with angle supports for cast iron fretted treads, often strengthened by castings for the risers (Figure 8.2b). In the USA external fire stairs are part of the vernacular architecture in the inner city (Figure 8.2c). They are sometimes considered as an integral element of the façade – referring back to the Hallidie Building, in Figure 4.32 – instead of becoming the visual accident that is commonplace. The American experience with fire escapes is more dramatic due to the hinged security steps for the lowermost flights. This device is utilized in the occasional stairs used for the diving stage of the Sherringhan Swimming Pool, detailed in Case study 14.5.

Fire escape construction is generally a mundane affair with column supports for framed landings and dog-leg flights

Figure 8.1e Addition of external steel stairs to Rochester Castle for visitors access (1990). (Courtesy of John Donat) (John Winter)

spanning as infill (Figures 8.2*d,e*). Detailing with structural sections requires skill to avoid clumsiness at the connecting points. Salvisberg's detailing is exemplary in the instance chosen where members are splayed or cut to ensure a neat geometry (Figure 8.2*e*). The construction is conventional in all other respects using 12.5 mm plate for strings and 105 mm × 105 mm tees with tubular rails. Case Study 14.4 illustrates how skilled detailing can transform the dullest building problem into the art of construction. Arne Jacobsen has solved the problem of an obstrusive external stair by encasing it in a cylinder of fire resisting glazing (Figure 8.2*f*).

Security can be effected by caging the fire stairs with exit gates at ground level (Figure 8.2*g*). The requirements of fire officers will include emergency lighting, fire resisting glazing to windows within a prescribed distance, and weather protection unless heated treads are provided.

In the UK external steel escape stairs are subject to licence and therefore regular inspections are needed to check on condition and repairs. Galvanizing the iron and steel components of the structure will prolong their life, delaying the paint treatment by ten or so years will increase the bond between the paint and the zinc coated surfaces.

Figure 8.2 Ladders and fire escape stairs
a Typical iron and stone steps in London areas

External fire escapes are feasible in the UK on the lines adopted by Gottfried Böhm at Rheinberg Town Hall (Figure 5.7). The advantage lies in the economy of construction as compared with the full incorporation of stairs within the building envelope. The Centre Pompidou is another complex where the principal movement areas are totally contained whilst emergency routes are largely external. Space is generally the problem but at Stockley Park there is an elegant solution for secondary external stairs (Figure 8.2b) and which can be bolted on to the façades as required.

8.3 Standard internal stairs

Fabrication methods devised for external stairs can be applied to internal stairs as with the Technical High School, Basle. (refer to Figure 8.2e). Open grid and chequer plate stairs are also used for industrial work. A higher standard of finish can be accommodated within metal trays or else laid on sheet steel formwork. The generic term is a 'folded sheet stair' welded to plate or tubular strings. It is this type of industrialized steel stair that is widely adopted in commercial buildings, particularly with steel frame construction. The advantages are linked to 'fast track' methods where the skeletal frame and stairs are erected in advance of floors and envelope walls, the folded sheet stair, even as formwork, provides access for the builders without the need for scaffolding towers. The pattern depends upon the finishes to be ultimately applied on site; Figures 8.3a, b, c depict the selection offered by fabricators and the range of finishes which can be achieved.

Individual trays can be fitted just before handover to save wear and tear in construction and can be used with stairs framed in timber or precast concrete (Figure 8.3d)

8.4 Spiral stairs

Steel spiral stairs are also the subject of industrial production and supplied as a package of parts. A typical assembly comprises a centre column and base plate, over which the cantilever treads or landings are sleeved, with cotter pins to secure each component in place

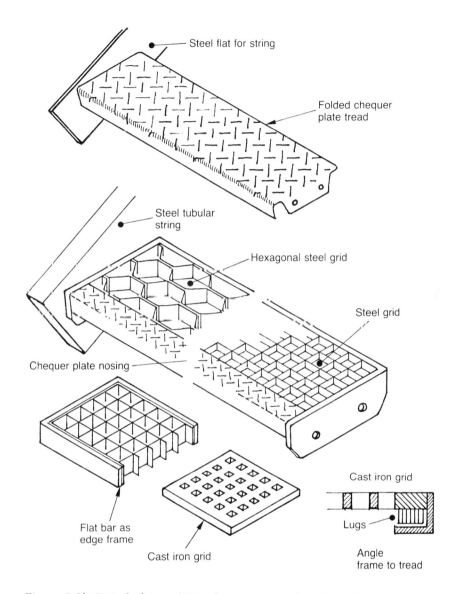

Figure 8.2b Details for traditional cast-iron and steel treads and risers

(Figure 8.4a). The handrail and rod balustrade stiffen the outside edge of the spiral. Welding the balustrade to guard rails can provide a structural lattice which acts as a supporting member, another method is to design a trussed balustrade (Figure 8.4b). The tread details follow the principles of 'steel trays' already discussed, with various materials acting as facing or infilling (refer back to Figure 8.3d). Designer spirals have reached beyond the limit of common sense in France with polished aluminium castings resembling giant teaspoons that are jettied off a spinal vertebra (Figure 8.4c). Patterns of nineteenth-century engineering are still manufactured, the developments in replica casting means that historic features can be matched. The spiral stairs and balconies at the Palm House, Kew are a case in point (refer to lead in Figure 8.1a), where the

Figure 8.2c Vernacular escape stairs in USA

Figure 8.2d Typical tubular framing to support landings of dog-leg stairs, offices at Stockley Park, 1989. (Arup Associates)

naturalistic forms still perfectly echo the natural plant material.

8.5 Special stairs

Welded plate formed into hollow tubular members gives the greatest freedom in design. The possibilities with cantilevers are more rewarding than standard spirals where bolted connections may loosen. The Dutch and French examples in Figures 8.5*a,b*, have a flair and verve that equals the finest work by Victor Horta. The imaginative use of curved rectangular tube in handrailing recreates the sinuous qualities associated with metals and is employed to maximum effect where the rails terminate as newels.

Public stairs within this idiom need to have a safe balustrade such as I.M. Pei's detail for the new entry below the Grand Louvre Pyramid (Case study 13.6). Curved sheets of toughened glass or acrylic (if permitted by fire codes) have inherent strength and sufficient rigidity to perform as a balustrade, if effectively clamped to the handrail and string (Figures 8.5*c, d*). The engineering to the Foster stairs and bridge at the Sainsbury Arts Centre, Norwich had to be tested with the built example on site to satisfy the local authority in 1977, though the construction principles are now fully accepted and marketed as standard balustrades.[1]

Glass and metal can also be combined in treads and risers – Jourdain's design at the Samaritaine Store is significant for its date 1901–10 (Figure 4.21). The theme recurs in the 1920s and 1930s with the glasscrete and steel grids used for landings and stairs at Maison de Verre (1927–32).

Stairs, Steps and Ramps   187

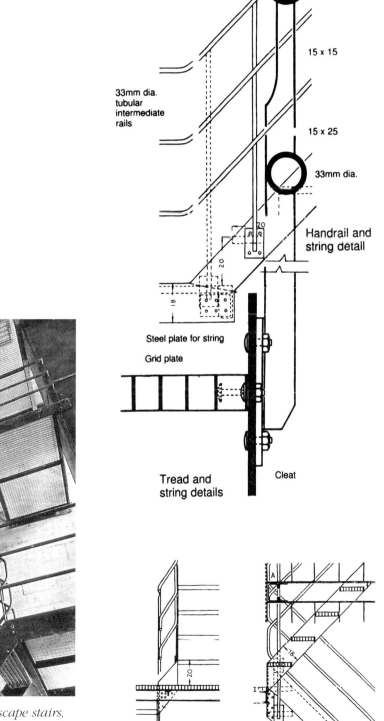

Figure 8.2e Typical steel framing: escape stairs, Technical High School, Basle, 1931. (Otto Salvisberg). (From Schuster, F., Treppen, Hoffman Verlag, 1949)

188 Stairs, Steps and Ramps

Figure 8.2f Minimal escape stairs with glass enclosure, 'Novo' factory, Copenhagen, 1961. (Arne Jacobsen) (Danish regulations permit spiral escapes)

Figure 8.2g Security enclosure to stairs. (Courtesy of Weland Grating UK Ltd)

Figure 8.2h External escape stairs, Apple Computer's Facility, Stockley Park, 1989. (Troughton McAslan). (See Case study 14.14a,c for plan and section)

Le Corbusier also experimented with 'Nevada' lenscrete to diffuse natural lighting to the stairwell in the Maison Clarté apartment building in Geneva (Figure 8.5e). The artificial lighting was arranged through a suspended conduit in the well serviced by naked light bulbs – replacement of the bulbs was effected by sliding the light fixture along its overhead track at the top floor and then walking down to each landing to change the bulb. The slender quality of steel profiles can reduce the impact of precast concrete or timber which may be selected for treads. Ladder

Stairs, Steps and Ramps 189

Figure 8.3 Folded steel stairs
a Providing builders access as skeletal frame is erected and then used permanently

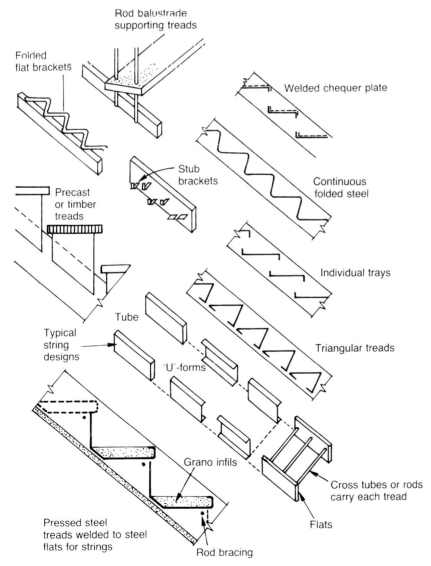

Figure 8.3b Profiles for folded sheet stairs or treads

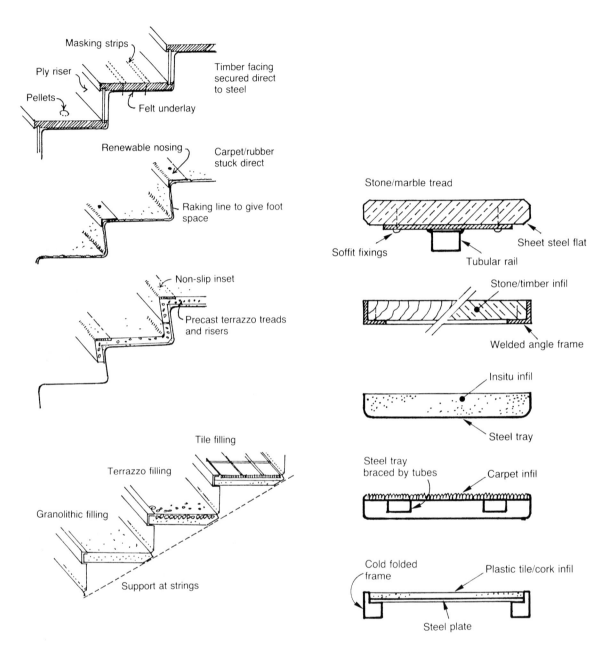

Figure 8.3c Finishes to folded plate   d Individual tread detailing

trusses with horizontal steel trellises will carry 'plank' treads, the upper line of the trusses forming a handrail. This motif was used frequently by Marcel Breuer for external stairs to reduce the impact or shadow from the construction interfering with the fenestration (Figure 8.5f).

Another approach is to separate the roles, with metal reserved for the tensile quality needed in handrailing juxtaposed with simple ships ladders for the wooden steps. Franco Albini's version demonstrated how practical utility becomes an art form when the sinuous qualities of

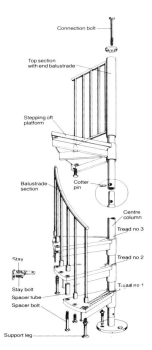

Figure 8.4 Spiral stairs
a Typical assembly of centre column with cantilever treads and landings (Courtesy of Weland Grating (UK) Ltd)

metal forms are fully expressed (Figure 8.5g).

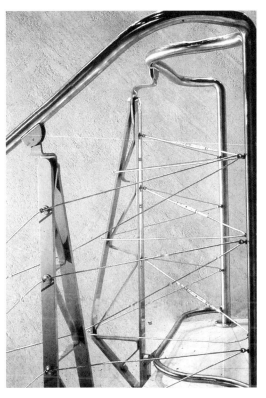

Figure 8.4b Lattice frame to balustrading detail of trussed balustrade to curving stairs at 'Legends' night club, London, 1987. (Eva Jiricna, engineers Dewhurst MacFarlane)

8.6 Suspended stairs

Suspended stairs have already been alluded to in Asplund's Courthouse (referring back to Figures 5.18a, b, c). The concept usually relies on individual rod suspension or else a structural balustrade which suspends the treads along the bottom chord (Figures 8.6a, b, c, d).

A lattice structure can be made by connecting handrail, balustrade and undercarriage brackets, various patents exist that apply to both curving and straight flights. Welding the balusters either as a horizontal frame (the hallmark of Marcel Breuer's designs) or as a trellis or vertical pattern are other alternatives. Suspension

Figure 8.4c Designer spirals beyond the limit of common-sense

192 Stairs, Steps and Ramps

Figure 8.5 Special stairs
a Welded plate to form hollow profiles for newel and for cantilever treads: accommodation stairs at City Hall, Arnhem, general view

Stairs, Steps and Ramps 193

b

c

Figure 8.5b Federation du Batiment, Paris, 1960s. (Badani, Foliasson, Kanjian and Roux-Dorlot)
c Spiral stairs, Sainsbury Arts Centre, Anglia University, 1977. (Foster Associates)

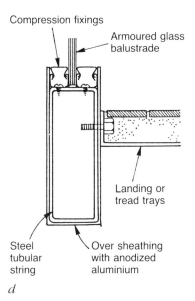

d

e

Figure 8.5d Construction detail of handrail balustrade and string, spiral stairs, Sainsbury Arts Centre
e 'Nevada' glass used in treads to Maison Clarté, Geneva, 1932 (Le Corbusier and Pierre Jeaneret)

194 Stairs, Steps and Ramps

Figure 8.5f Ladder trusses carrying treads. Einstein Village, Princeton, USA, 1957. (Marcel Breuer)

wires can be used to secure wooden treads, one of the most minimal solutions being devised by Peter Moro at the former Hille showroom. The interior is now destroyed but is recalled to demonstrate the thesis 'least is most' with this example of slender detail (Figure 8.6d). The most interesting combination of trussed suspensions has been developed by Eva Jiricna and her engineer consultants. One of the best crafted stairs is the triple flight incorporated into the Joseph Store, Sloane Street (Figures 8.7a, b, c). The

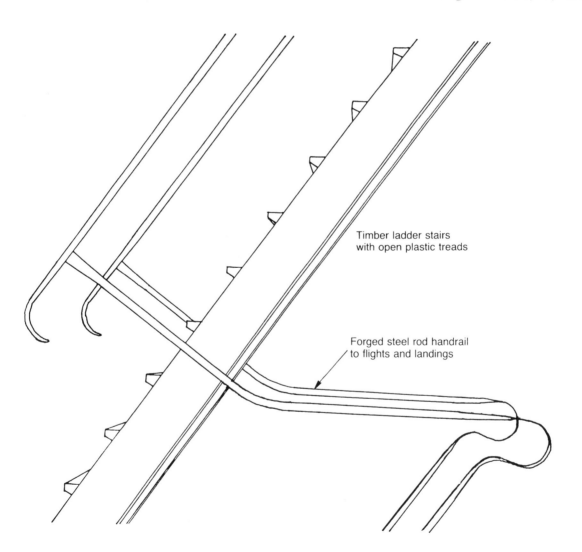

Figure 8.5g Art forms with sinuous handrailing. Cottage stairs at Somma, Italy, 1960s. (Franco Albini)

*Figure 8.6 Suspended stairs
a Patented balustrade brackets (Courtesy of Iconi)*

sculptural effect of this stair is closer to furniture with the engineering honed to an absolute minimum. The lightness of this eye catcher is enhanced by glass treads and balustrading whilst the structure itself is an assembly of stretched cables, stainless rods and connectors providing a spider's web of supporting members. The art of constructing stairs when this level of skill is required is a combination of architectural and engineering input hence the multiple credits of Eva Jiricna and Matthew Wells, now a partner with the consulting engineers Barton and Wells.

The flying ramps at the Imagination Building or the hovering landings of John Young's flat are in the same league. Designers often return to themes explored by others in times past, James Holland and Peter Chamberlain's detailed concept for the lattice stairs for the British Industries Fair (Figure 8.7d) is an original British work for lightly suspended structures.

8.7 Metal balustrading

The choice of bronze or steel for balustrading is dictated by the engineering. Problems with the proven strength of bronze imply that bronze, like aluminium, is often utilized as a sacrificial surface over steel cores in structuring balustrades. The constructional restraints are governed in the UK by Codes of Practice[2] and are concerned with lateral strength and fixings. This is a separate issue to that raised by the Jiricna–Wells suspended stairs where the total ensemble of balustrading and stair framing have to be assessed as structural engineering.

The traditional concept can be compared to a framework infill, comprising the following elements and illustrated in Figures 8.8a, b, c, d, e, f and g.

- *Vertical Members* termed standards; provide the main support.
- *Rails* termed handrails and core rails, the latter making a sub frame (if needed) for the infilling members.
- *Balustrades* formed with bars, mesh or sheet materials.
- *Connections* welding, set screws, sleeves and cleated brackets.

The assembled framework is connected together in panels by set screws and on-site welding, the supporting strings can be timber, steel (plate or tubular) or reinforced concrete (either *in situ* or precast) (refer to Figure 8.8g).

The geometry of the assembly is critical to the overall stability, long runs to landings and straight flights will need heavier standards or stays. Shorter flights with dog-leg or triple-turn layouts provide sufficient return lengths to stiffen the framing.

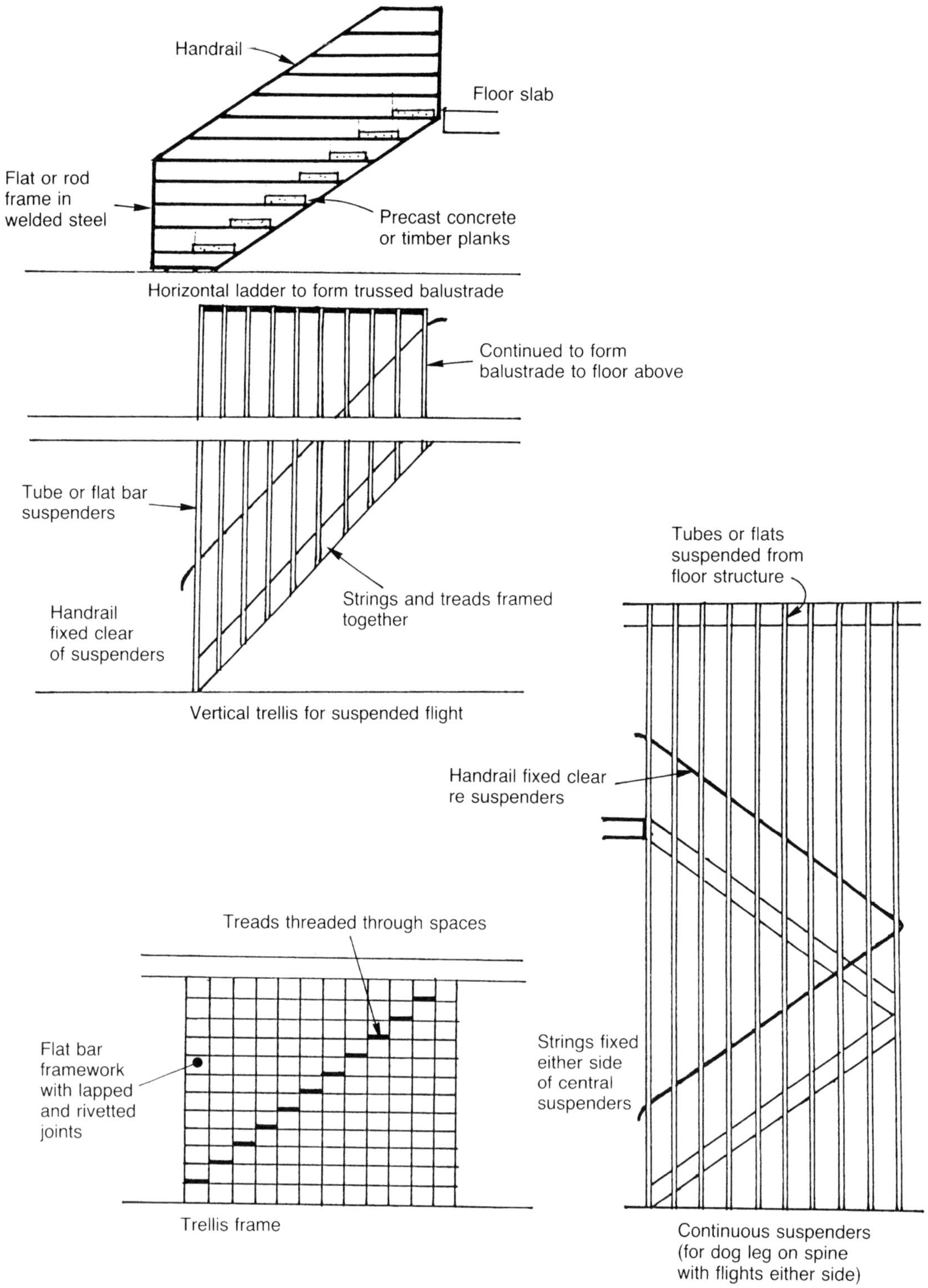

Figure 8.6b Balustrade trusses

Stairs, Steps and Ramps 197

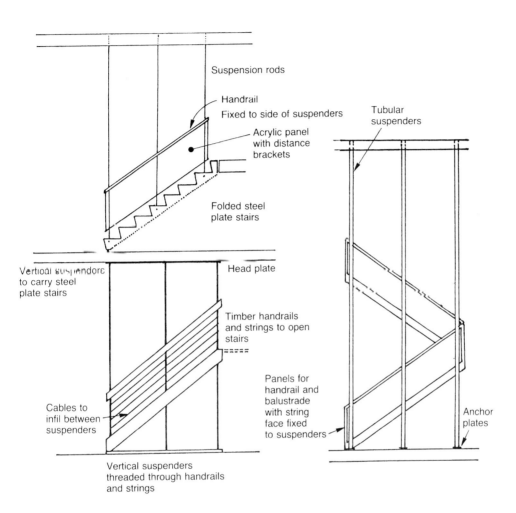

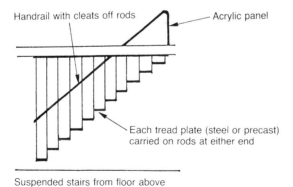

Figure 8.6c Suspended stairs

198 Stairs, Steps and Ramps

Figure 8.6d Cable suspension, Former Hille Showroom, London, 1963. (Peter Moro)

Circular balustrading is also stiffer than straight runs.

Base fixings are another factor, direct vertical fixings have less torque than side mounted standards. The advantage of the former stems from the face-to-face bolting that can be obtained from base plates to the stair or landing structure, with loading in line with the standards. The actual connections vary with the stairs, refer to Figure 8.8 for a range of alternative details.

Side fixings into the edge of treads or landings save space since Building

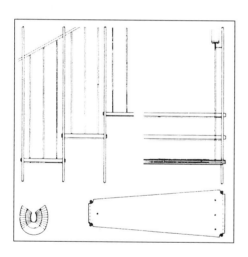
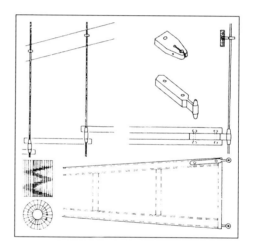
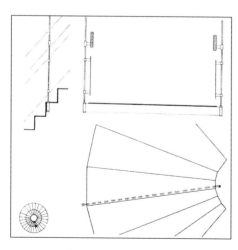
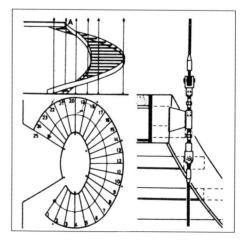

Figure 8.6e Details of suspender connections for spiral stairs. (From Pracht, K., Treppen, Deutsche Verlags-Anstalt, 1986)

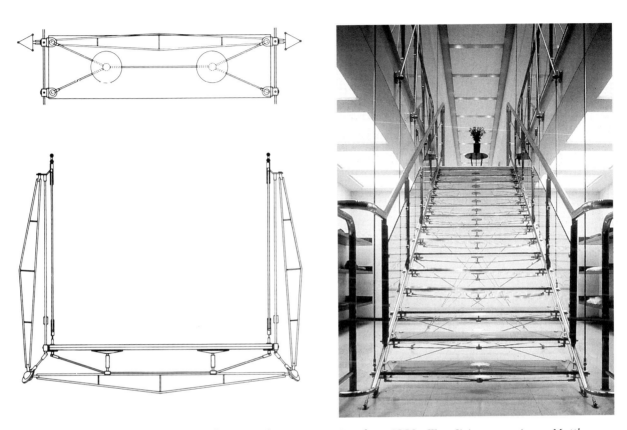

Figure 8.7 Suspended stairs, Joseph store, Sloane Street, London, 1989. (Eva Jiricna, engineer Matthew Wells of Barton and Wells)
a Details: plan and section through typical tread *b General view. (See also Figure 4.28)*

Codes in the UK and the USA permit the required stair width to be measured to the edge of the handrail. The actual girth of tread is therefore minimal as compared with base mounted balustrades which can add 75 mm to the width of stair or landing construction at balustrade locations. Side fixings do not necessarily save space since the well size has to be sufficient to accommodate the 'L'-shaped standards apart from adequate dimensions if wreathings are employed. Balustrades set clear of tread and landing surfaces have the crucial advantage that floor coverings are not perforated which makes maintenance and replacement an easier proposition.

A compromise solution is to devise a string component for both treads and landings which contains vertical fixings for the balustrading and horizontal connections to the surrounding structure. Traditional detailing involves hardwood sections (Figure 8.8g) but tubular steel sections or pre-cast concrete will serve equally well, the latter has the advantage that factory made finishes such as terrazzo or tile can be applied to give a high quality setting for the balustrade. The arrangement also means that the minimal structural width can be provided for treads, etc. whilst the handrail/standard/string component occupies a zone within the well. Straight string features that project above

Figure 8.7c Detail of structure

nosing lines to stairs and landings also facilitate maintenance whereby cleaning operations do not spill dirt down the staircase well.

The 'guarding' rules for balustrades to stairs and landings under the National Building Regulations are explained in Chapter 11 Section 2 with illustrations in Figure 11.2*a*. The worst situation occurs where multi-occupancy is planned and the designer is faced with a variety of conflicting advice under the Regulations as to the ideal height for the handrail. Aalto resolved this dilemma by running handrails in duplicate, one for the adults and one for the children.

Another sensible solution is to choose a standard height for the principal balustrade element, be it railings or walling and then mount the handrail separately

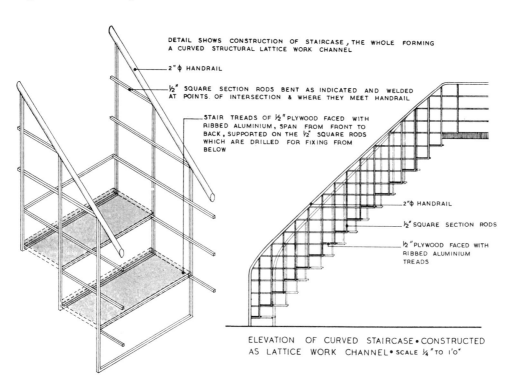

Figure 8.7d Lattice stair framing for Lafarge Stand, Building Exhibition, 1949. (Designers James Holland and Peter Chamberlain)

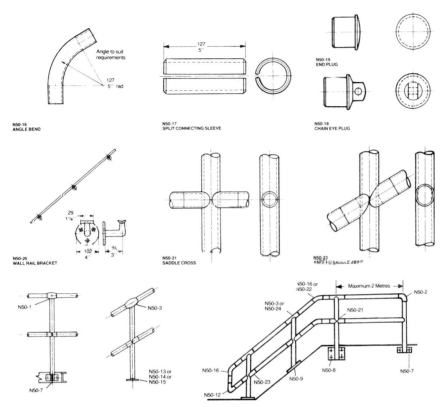

*Figure 8.8 Traditional metal balustrading
a Industrial balustrading. (Courtesy of Norton Engineering Alloys Co. Ltd.)*

to follow the mandate of the regulations. In this method the geometry of the spandrel treatment (perforated or solid) can at least match the stair geometry.

Aluminium or galvanized steel industrial guard-railing is another 'off the peg' balustrading that is effective in certain categories of work – safety barriers for maintenance stairs and accessways, even external works. The meccano like components comprise universal sockets and tubes held together by inset screws. The appearance with 'arthritic' joints may not be to everyone's taste, mesh panels will need to be fitted where children are at risk.

A non-traditional concept is to place the balustrade assembly as a vertical infilling to a minimal stair well. The panels are supported by newel posts and the handrailing bracketed to provide a continuous hand hold as in Figure 9.5f. Stretched cable balustrades is another novel adaption of marine detailing borrowed for buildings. The rhythmic variation of metal balusters is perhaps the most memorable image (variations are given in Figures 8.9a,b,c,d). The concluding illustration (Figure 8.10) is a typical Miesian staircase – the basement flight in the Neue National Galerie, Berlin. The arrangement follows that seen in other buildings by Mies van der Rohe and must follow the office routine in detailing with minimal square tube. The balustrades have another point in common, they all wobble at the last newel post.

202 Stairs, Steps and Ramps

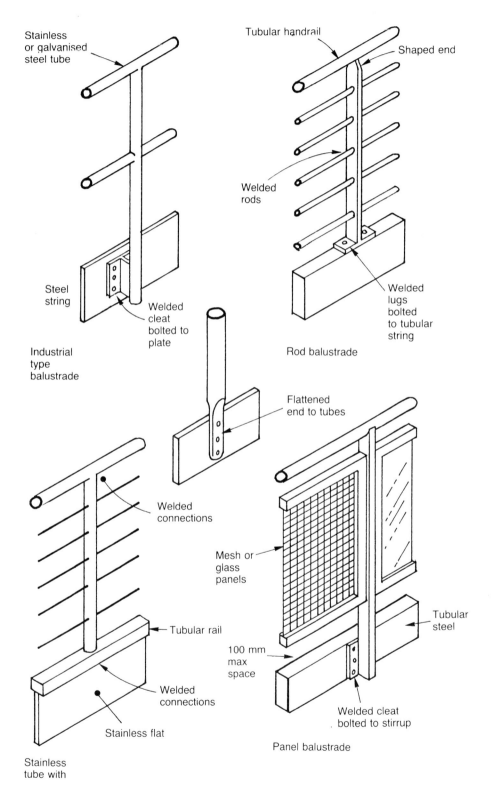

Figure 8.8b Balustrade panels

Stairs, Steps and Ramps 203

Figure 8.8c Mesh panels. (Courtesy of Aidrail Ltd)

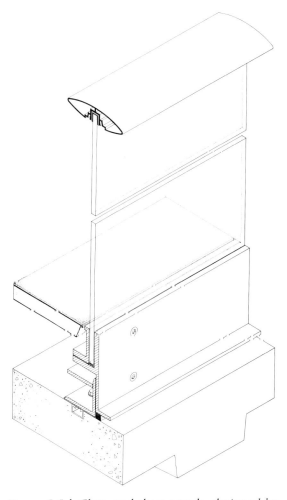

Figure 8.8d Glass and sheet panels, designed by Foster Associates

Figure 8.8e Horizontal rail. (Courtesy of Aidrail Ltd)

204 Stairs, Steps and Ramps

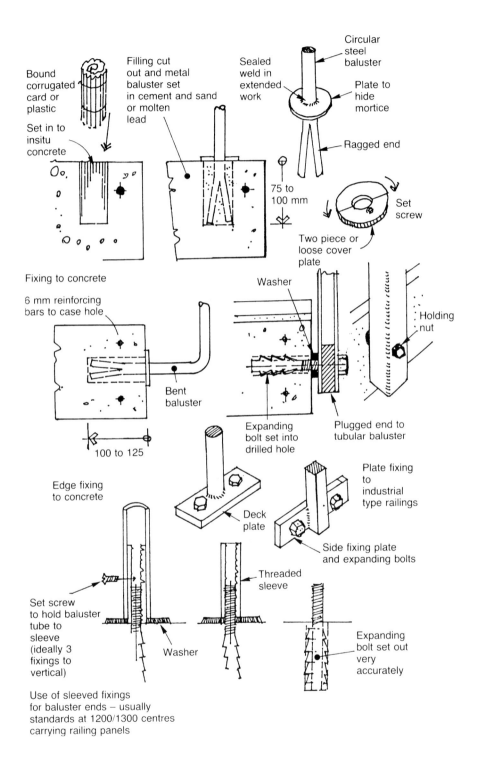

Figure 8.8f Connections (welding, set screws, sleeves and cleated brackets)

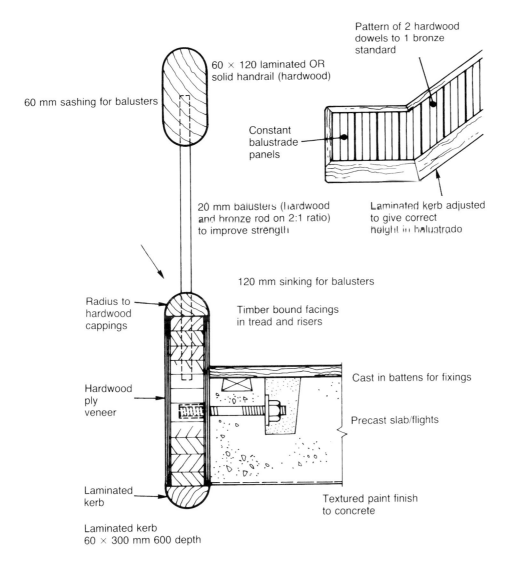

Figure 8.8g String to balustrade details (timber, steel, concrete)

206 Stairs, Steps and Ramps

a

b

Figure 8.9 Unusual construction
a Wrought iron loops, John Soane Museum, London, 1790
b Brass frame, with handrail, enamelled standards, wires as infill and capping to string, Åarhus City Hall, 1943. (Arne Jacobsen)

c

d

Figure 8.9c Trussed balustrade with steel handrail, verticals and folded steel treads, Havas Conseil, Paris, 1960s
d Rhythmic decoration for Kedleston Stairs, 1790s. (Robert Adam)

Figure 8.10 Miesian standard stairs, Neue National Galerie, Berlin, 1965–68.

References

[1] There are a number of handrailing specialists who fabricate standard assemblies which combine steel tubular rails with glass and/or acrylic. Typical products in the UK are (Figure 7.8*f*) 'Hewi' (metal and glass from Hewi (UK) Ltd.); (Figure 8.8*a*) 'Neaco and Nearail' (metal and glass from Norton Engineering Alloys Ltd); (Figure 8.8*c,e*) glass, mesh and rail systems from Aidrail Ltd.

[2] British Standards Code 6180 (1982) lists strengths required, e.g. sufficient to resist a horizontal force for private or common stairs and 0.74 kN for the remaining categories of stairs under UK regulations. (See Figure 11.2*a*).

9 Concrete stairs

9.1 Early forms

The earliest forms of concrete stairs are Roman where the concrete was placed over brick vaulting to support ramps or steps such as in the construction of the Colosseum (Figure 9.1a). The introduction of filler joist flooring in the eighteenth century with wrought-iron beams infilled by brick arches was adapted to staircases cantilevered off masonry walls (Figure 9.1b). The tradition continues wherever brick vaulting techniques are still employed. Refer back to Figure 2.11 where the Spanish design, although eighteenth-century in origin, can be seen being constructed today around much of the Mediterranean.

The early use of concrete in the twentieth century was found in composite forms of building. In the case of stairs, steel joists often formed the stringers to reinforced

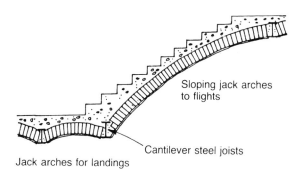

Figure 9.1b Filler joist stairs constructed with steel beams and brick vaults overlaid with rendering

concrete infilling (Figure 9.1c). These ideas arose from the framing requirements of Building Bye-laws at the turn of the century where the notion of framing meant 'frame and panel'. The first experiments with reinforced concrete led to similar conclusions until designers were confident in handling the new material in terms of monolythic planes, whether floor, stair or wall slabs. The Perret archives show views of the early reinforced concrete stairs completely bare of decoration or metalwork and appearing as progressive as Nervi. A visit to the interiors at 22 bis rue Franklin reveals the conventional fitting out which belies the core to the design. Nervi on the other hand was allowed to leave the remarkable double spiral stairs expressed as concrete structure (Figure 9.1d), though he was obliged to add a classical colonnade to the entry elevation by the Fascist authorities in Florence. Nervi's avant-garde approach to reinforcement design is not without problems – loss of cover to the exposed

Figure 9.1a Brick vaults which supported steps at the Colosseum, AD 72–82

210 Stairs, Steps and Ramps

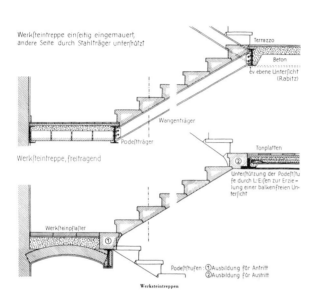

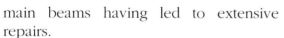

Figure 9.1c Textbook example of early reinforced concrete stairs using composite construction with steelwork

Figure 9.1d Confident use of reinforced concrete structure in stairs, Florence Stadium, 1932. (Engineer Pier Luigi Nervi)

main beams having led to extensive repairs.

Both Frank Lloyd Wright and Le Corbusier were enthusiastic exponents of reinforced concrete and realized stair designs that captured the plastic qualities of the material. The long curving ramp at the Guggenheim Museum (Figure 2.20a,b) is perhaps the most celebrated cantilevered form. Le Corbusier's designs from the heroic period (1920–30) transformed ramps and stairs into sculptured volumes that interpenetrated the floor spaces of his major buildings. The poetic geometry given to the circulation elements within the Villa Savoye (referring back to Figure 3.13) is amplified in the palatial volumes made for grand projects such as The League of Nations or The Centrosoyus Building, Moscow. Refer back to Figures 5.23a and b and accompanying text for a fuller description. The staircase structure and its relation to the plan in the Goldfinger House in Hampstead is a microcosm of the Corbusian principles, revealing the key to employing monolythic concrete to form both column and curving slab within a single stair (a view is given in Figure 3.14 and a structural diagram in Figure 9.1e).

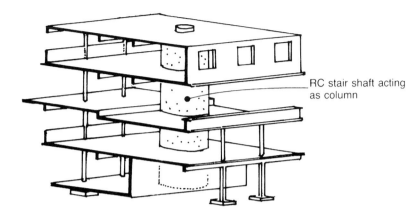

Figure 9.1e Structural role of stair, Goldfinger House, Willow Road, Hampstead, 1937. (Ernö Goldfinger)

9.2 Developed forms of reinforced concrete stairs

(Figures 9.2a, b, c, d)

In situ reinforced concrete has the same limitation of span to thickness as floor slabs so that single span flights are limited to about 4500 mm unless downstand spine beams or upstand balustrade beams are used. The spine beam version is usually pre-cast and implies pre-cast or metal framed cantilever treads.

Dog-leg stairs are usually carried on cross beams at half landings or else on a thickened landing slab. Balustrade beams convert slab forms into slabs with twin upstand components which permit clear spans floor-to-floor with dramatic shapes, as exemplified in the foyer to Sir Denys Lasdun's National Theatre (Figure 2.27). Lasdun's use of concrete surfaces and texture celebrates the sculptural possibilities of a material that in monolithic construction can perform many roles. The foyer lighting enhances the surface treatment of board marked concrete and emphasizes the importance of the stair shafts rising through the extensive volume (Figure 9.2f). The concrete profile echoes the tread and rise relationship while the stainless handrailing (bracketed off the solid balustrade) follows the awkward dictates of bye-laws without spoiling the overall form (resembling Figure 9.2d).

A more expressionist approach is seen in the Goetheanum, Dornach (Figure 9.2g) where Rudulph Steiner's architect seeks an atmosphere closer to the interiors of the film *The Cabinet of Dr Caligari*[1] (Figure 9.2h). The concluding concrete work in this preamble has the essence of Henry Moore where the space between the solid forms are as important as the forms themselves. The designers Novello and Lange have taken a continuous curving slab and placed this eccentrically within a smaller 'eye' in the floor above. The attraction relies upon the combination of the rebated stepped profile with the varying curves of ceiling plane and handrailing. The carpet runs as a continuous band of colour (Figure 9.2i).

Repetition in concrete stairs favours pre-casting the components, it is difficult to generalize but with stair costs in excess of £150,000 it is worth considering precast manufacture that will simplify details into variants of Figure 9.2b and c.

212 Stairs, Steps and Ramps

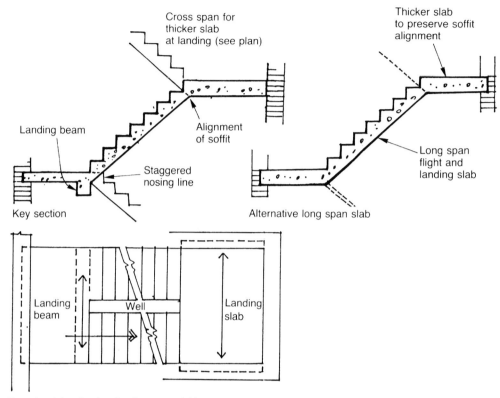

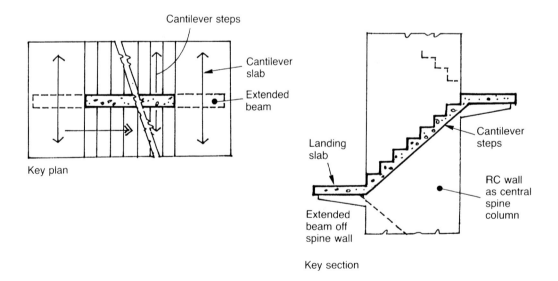

Figure 9.2 Developed forms of reinforced concrete stairs
a Simple RC slabs (with and without landing beams)

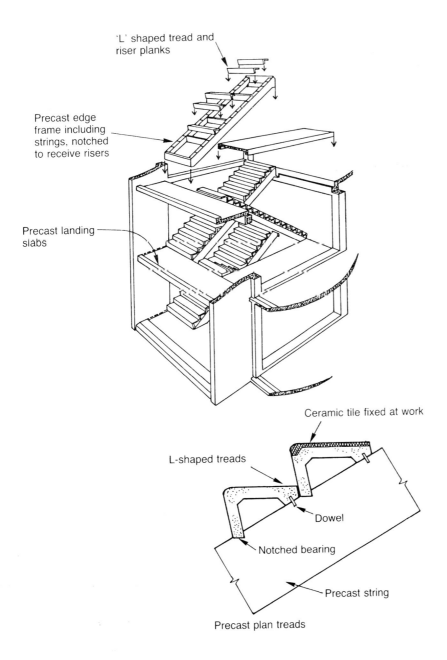

Figure 9.2b Precast treads and beams

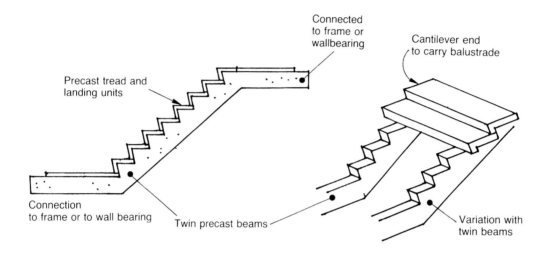
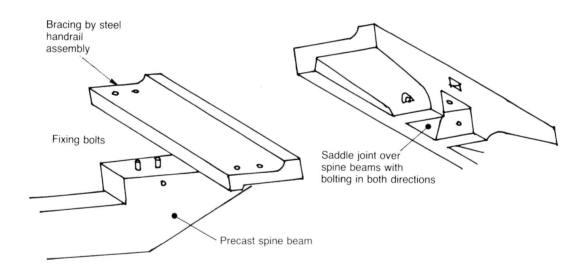

Figure 9.2c Spine beam support (with pre-cast treads)

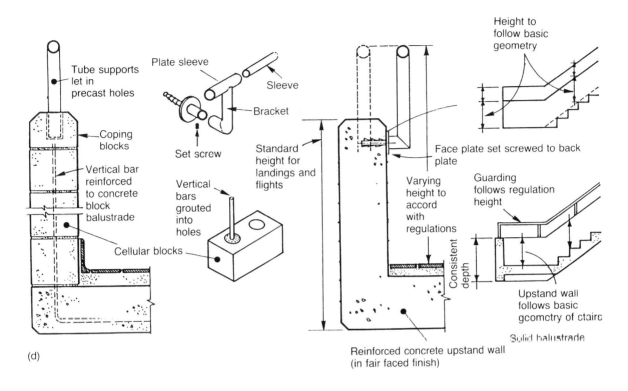

Figure 9.2d Balustrade upstand beams

Pre-cast stairs can be made in site factories (Figure 9.3a) and completed at this stage with a range of finishes similar to that achieved for cantilever treads (Figures 9.3d, e, f). Spine beam stairs are more efficient for long straight (Figure 9.3b, c) or curved flights due to the lower weight per span compared with solid slabs. Pre-cast assembly increasingly depends upon steel bolt and plate fastenings which demands the accuracy and connection methods associated with structural steelwork. Pre-cast reinforced treads are often cast with the finishes intregal with the component, terrazzo can be used as a trowel coat within the mould and as a wearing coat to the tread, the final grinding and finishing follows the curing stage. A spine wall can serve as a dog-leg column with cantilever action in the treads, by contrast the shaft walls can have cantilever steps leaving the well open for a metal balustrade.

The profile chosen by Erskine at Clare Hall, Cambridge, allows the carpet to be wrapped around the concrete treads, the rounded corners preventing wear and permitting the carpet to be turned from time to time. The reinforcing is tied into the backing wall which forms the shell-like enclosure (Figure 9.2e).

9.3 Precast spiral stairs

The general arrangement of pre-cast spiral stairs is similar to steel spirals with tread units threaded over a central tubular column. The concrete profiles are more robust, while finishes like terrazzo or tooled concrete (with stone aggregates

216 Stairs, Steps and Ramps

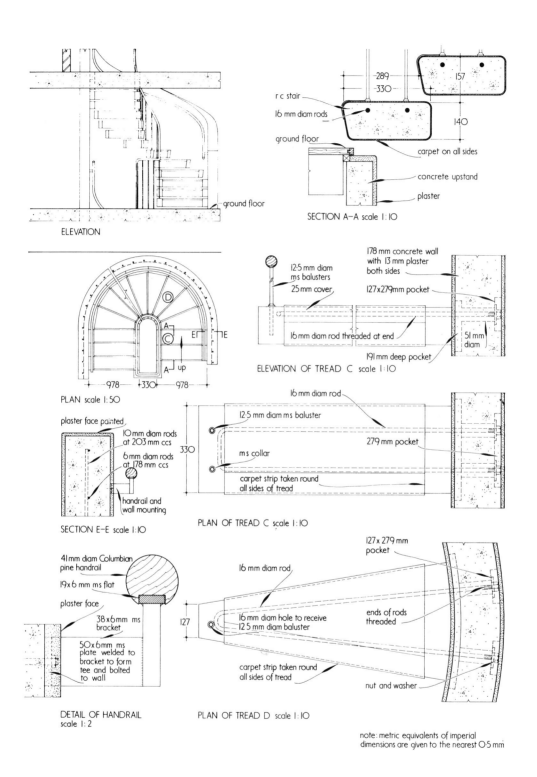

Figure 9.2e Cantilever concrete stairs
Clare Hall Cambridge, 1969. (Ralph Erskine, in association with Twist and Whitley)

Stairs, Steps and Ramps 217

Figure 9.2f Cast in situ *concrete work to stair shafts, Royal National Theatre, 1967–76. (Denys Lasdun & Partners)*

Figure 9.2h Interiors from film 'The Cabinet of Dr Caligari' *(Germany, 1919) directed by Robert Wiene. (Designers Walther Rohrig, Herman Warm and Walther Reimann)*

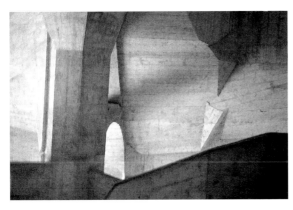

Figure 9.2g Cave like in situ *forms, stairways Goetheanum, Dornach, 1924–28. (Hermann Ranzenberger)*

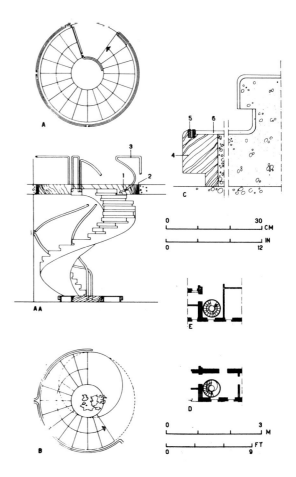

Figure 9.2i Concrete spiral stair, Villa at Bullono. (Novello and Lange, Engineers Bergoglio and Mutti)

*Figure 9.3 Pre-cast stairs
a Site factory at Wates Housing Development*

Figure 9.3b Precast beam stair prepared for tread fixing

Figure 9.3c Completed work with treads and balusters in place

or coloured cement) can produce surfaces as good as natural stonework. They look well externally (Figures 9.4a, b, c, d).

A novel interior application for spiral cellars was sold in France for the 'Cold War' years. The kit designed for sinking below the home comprised shaft walls, base and a set of perforated spiral treads cast with circular bottle sized holes (running front to back). One assumes however that the 1976 Mersault and corkscrew, with Napolean brandy would be kept within the lowermost steps in case of the four minute warning (Figure 9.4e).

9.4 Detailing of well ends and balustrades

The solid profile of *in situ* concrete stairs draws attention to the riser and well end relationship. The appearance of sloping soffits and their bisection is of critical importance (Figure 9.5a, b) unless trimming beams or thicker slabs across the landing hide the junctions (Figure 9.5c). This may however appear clumsy. The alignments are compounded by the handrail-to-riser geometry, where 'slipping' the riser plane by one full step will improve matters (Figure 9.5b). It will be noted that the handrailing bisects in the same plane as the soffit and that no jump is incurred for the wreathing. In other words, the alignment gives 'slope to horizontal to slope' in the handrailing at well ends. By comparison, Figure 9.5a gives a vertical jump that is visually awkward.

Lengthening the well is another possibility with the use of a semicircle to set out handrail and well end. A well width equal to the tread will give a lift to the handrail

Stairs, Steps and Ramps 219

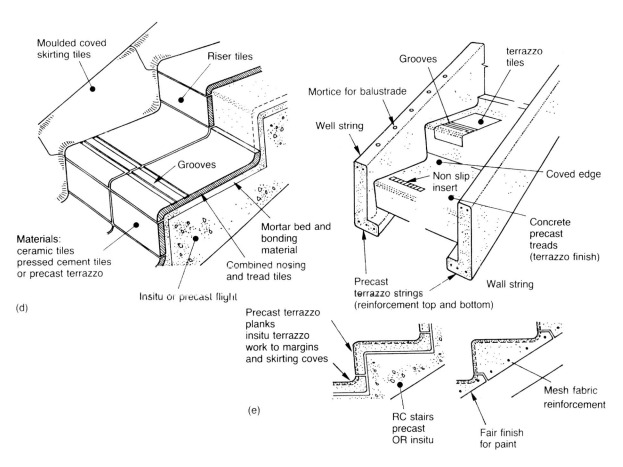

Figure 9.3d Tile finished treads *e Terrazzo finish*

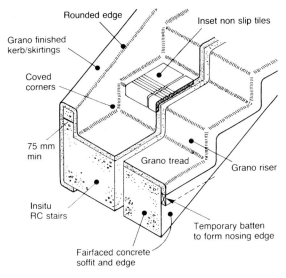

(f) Use of tile and insitu finishes either grano or terrazzo

Figure 9.3f Combined finishes to treads

220 Stairs, Steps and Ramps

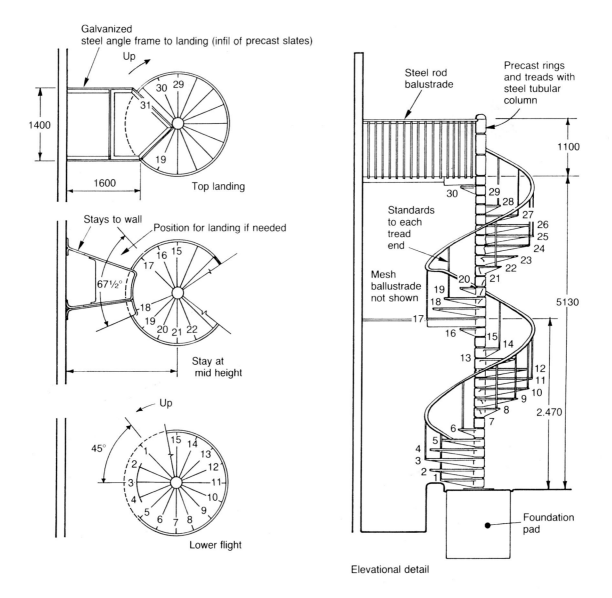

Figure 9.4 Pre-cast spiral stairs
a Typical components using pre-cast treads and steel tubular column. (Courtesy of Cornish Spiral Stairs Ltd)

Figure 9.4b Finished tread with terrazzo surface of different colours including non-slip insets

Figure 9.4d Context of artificial stone stairs set against stone work, Carcassone, 1870s

Figure 9.4c Artificial stone finish to spiral stair. (Courtesy of Cornish Spiral Stairs Ltd)

in sympathy with the pitch of the flight (Figure 9.5*d*). A compromise which saves a few centimetres of space is the arrangement in Figure 9.5*e*, where the riser plane is slipped by half a step. This might please valuation surveyors – referring back to Chapter 4, Section 1. There are limitations however on aesthetic issues particularly where a matching balustrade pattern is needed either side of the stairwell. In these circumstances non-conventional solutions such as solid fender walls or face mounted railings will have to be explored (Figure 9.5*f*).

Generous layouts as in Figures 9.5*b* and *d* give better value in visual terms for both balustrade and well shape despite adding to the unlettable area a sliver of space 250 mm (average tread) × stairhall width. The pre-war taste of Franz Schuster is totally persuasive in this respect.

Metal balustrades to concrete stairs follow the principles laid down in Chapter 8, the differences arise if reinforced concrete

222 Stairs, Steps and Ramps

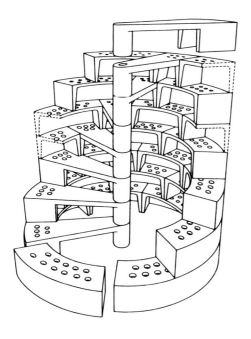 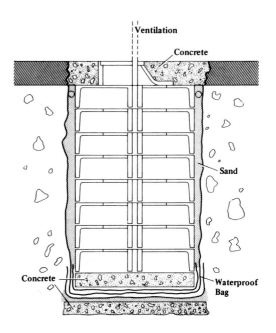

Figure 9.4e French cellar steps doubling as an air-raid shelter

fender or spine walls are involved. The advantages of the latter rest with less metal work to install and maintain. Figure 9.5g demonstrates Aalto's detailing at the Paimio Sanitorium as a superb example. The photograph was taken in 1968 when the original finishes were still intact. Another version of the concept could be made by using pre-cast terrazzo to the strings to receive the railings, with the flights also pre-cast with terrazzo tile finishes (Figure 9.5h).

Visitors to the former empire of the Soviet Union will have noticed the degree of industrialization in concrete stairs, with identical production spanning from Leningrad (as it was called) to Alma Alta on the Chinese border. The lengths and sizes of concrete were totally standardized, resulting in the site engineers' inability to fit the building to the units. A typical experience is illustrated in Figure 9.5i where the floor has a crash junction, with a tiny 75 mm riser, to be countered by a heftier step than normal where the staircase meets the floor above.

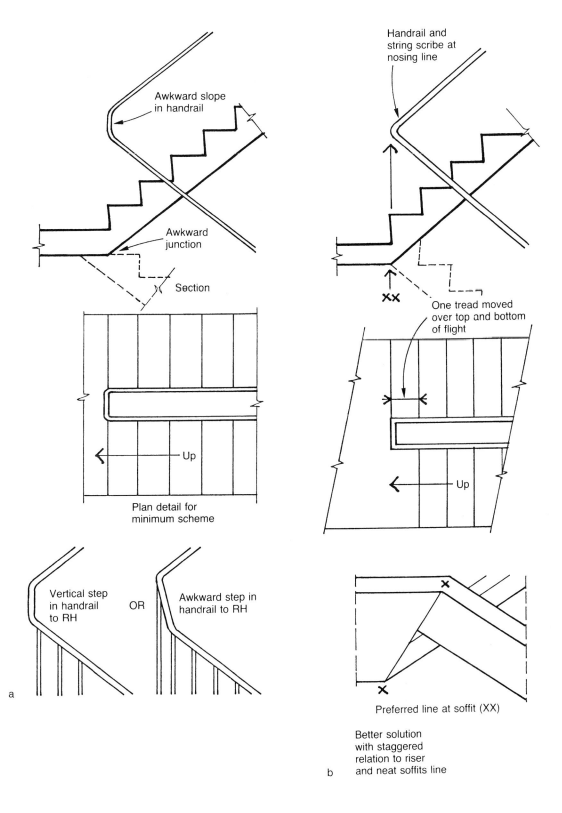

Figure 9.5 Detailing of well ends
a and b Sloping soffits and their bisection, good and bad solutions

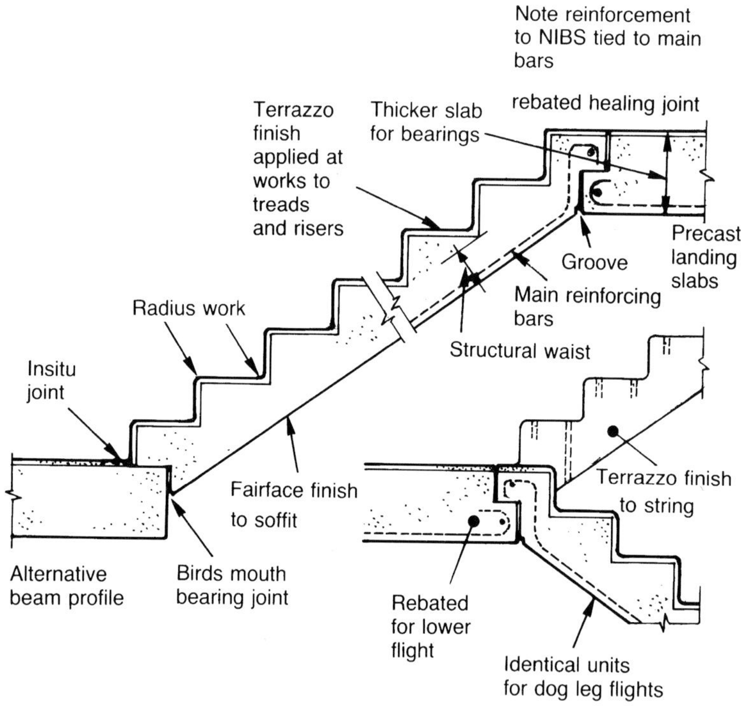

224 Stairs, Steps and Ramps

Figure 9.5c Beam or thick landing slab to mask intersection lines

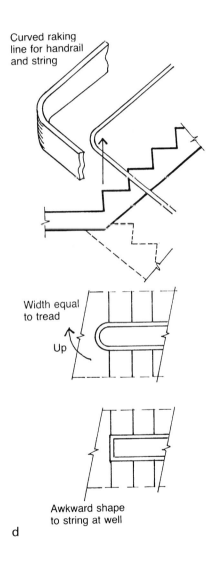
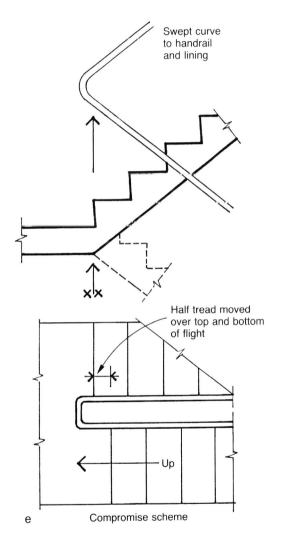

Figure 9.5d Lengthening the well.   e Half-step relation accross well

226 Stairs, Steps and Ramps

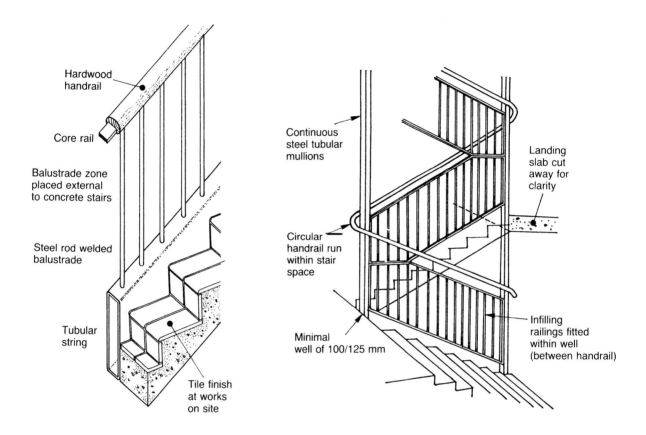

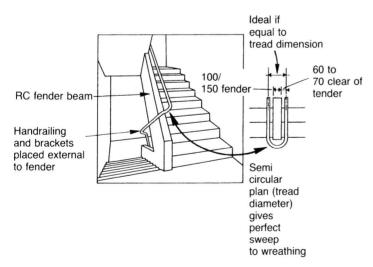

Figure 9.5f Face mounted railings

Figure 9.5g Fender walls with least balustrading, Paimio Sanitorium, Finland, 1933. (Alvar Aalto)

Figure 9.5i. Pre-cast work that never fits (example from former USSR, 1987)

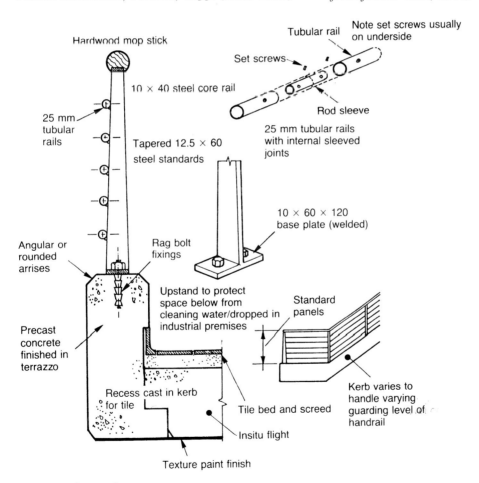

Figure 9.5h Fender in pre-cast terrazzo

References

[1] Details of *The Cabinet of Dr. Caligari* may be found in *The Film till Now*, Paul Roth, Vision Press Ltd., 1949.

10 Stonework and composite construction

Composite stair construction has already been mentioned with regard to steel and glass alongside steel and concrete, or steel working with timber. Traditional building methods relied on stonework slabs or vaulting or on stone cantilevers built in conjunction with iron balustrades.

10.1 Stonework in general terms

The historic review in Chapter 2 touched on stonework construction although it was primarily concerned with the systematic comparison of stairs. It is useful to list the types of construction and to cross reference these to the buildings illustrated in Chapter 2 alongside symbolic sketches in Figures 10.1*a–e* giving the principles involved.

- Stone slabs spanning enclosing walls, e.g. steps to roof terraces at Dendera (Figure 10.1*a*, and compare Figure 2.4*b*).
- Stone blocks as a stylobate, namely stonework forming a stepped plinth, e.g. the steps to the Propylaea and Nike Apterous (Figure 10.1*b*, and compare Figures 2.5 and 2.6).
- Stone vaults, either as a sloping barrel vault carried on walls or as cross vaults supported on columns, e.g. Palazzo Municipio, Genoa or double circular stairs at Blois and Chambord (Figure 10.1*c*, and compare Figures 2.10, 2.17 and 2.18).
- Turret stairs, with treads used as bonding stones between two independent towers, e.g. west towers, Sagrada Familia, Barcelona (Figure 10.1*d*, and compare Figure 2.16).
- Turret stairs, with treads tapered and built up as a central newel or pier with tread ends built into the enclosing wall, e.g. the stairs within corner piers at Chiswick House, London (Figure 10.1*e* and compare Figure 2.12), or else treads cantilevered out towards an open well (Figure 10.1*f*).

10.2 Turret and cantilevered stairs

One source for the dramatic structures of turret and cantilevered stairs is the influential writing of Andrea Palladio. The text from 1570 *I Quattro Libri del Architettura* (*The Four Books of Architecture*) devotes significant illustrations and text to the advantages of open well stairs constructed with stone cantilevers. Two of the designs, Palladio's staircase in the Carita Monastery, Venice and for the Palazzo Capra, Vicenza (Figures 10.2*a, b*) were both seen by Inigo Jones. His copy of *Quattro Libri* has a Jonesean sketch of the Palladian stairway at Carita and carried notes on the oval stairs at Palazzo Capra

Stairs, Steps and Ramps   229

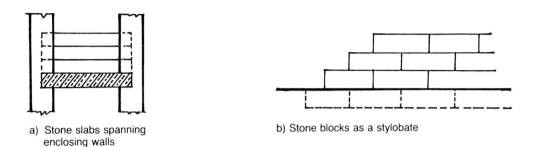

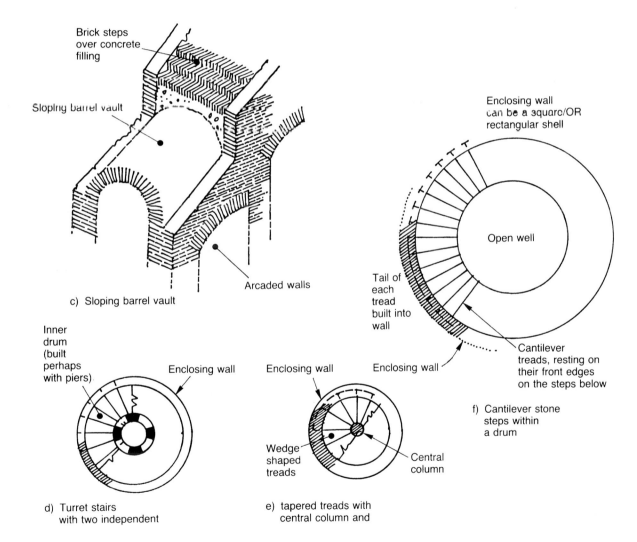

Figure 10.1 Principles for stonework stairs
a Stone slabs spanning enclosing walls
c Sloping barrel vault or sloping cross vault
e Tapered treads with central column and enclosing wall

b Stone blocks as a stylobate
d Turret stairs with two independent towers
f Cantilevered stone stairs within a drum

230 Stairs, Steps and Ramps

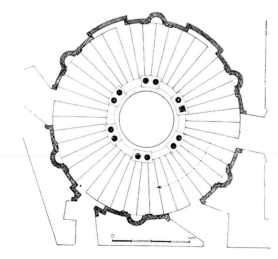

Figure 10.1g(ii) Plan of spiral stairs, Villa Farnese. (Photo and plan from Schuster, F., Treppen, Hoffman Verlag, 1949)

Stairs, Steps and Ramps 231

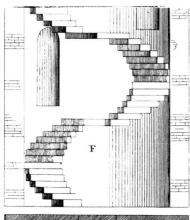

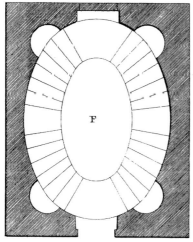

Figure 10.2 Turret and cantilever stairs a Plan and section of stairs at the Carita Monastery, Venice, 1560–61. (Palladio)

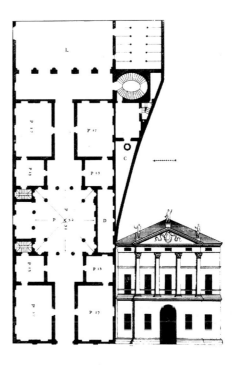

Figure 10.2b Plan of Palazzo Capra, Vicenza (from Quattro Libri, *1570). (Palladio)*

as follows: 'The Stairs are Parted with a raille of iorne so that you see through the parts underneath and so every waie.' Palladio clearly prefers open well spiral stairs for he says of them in *Quattro Libri* 'They succeed very well that are void in the middle, because they can have light from above, and those that are at the top of stairs see all those that come up or begin to ascend, and are likewise seen by them'.

Inigo Jones certainly drew upon the inspiration of Andrea Pallidio in the 'Tulip' staircase in the Queen's House Greenwich (Figures 10.2c, d). A technical refinement occurred whereby the treads were rebated at the front bearing edge on the step below. The stone mason responsible was Nicholas Stone, later to become the architect associated with the entry screen to the Oxford physic garden. The rebating prevents slipping in the building stage and improved the bearing from one tread to the next as compared with the simple bed joint illustrated by Palladio (refer back to Figure 10.2a). Other notable English stairs built on the cantilever principle in the seventeenth or early eighteenth century are the open spirals at the Monument, City of London (Figure 10.2e) and the great stairs of Chatsworth (Figures 10.2f, g). Wren and Hooke deployed at the Monument a moulded riser, which features in full at the tapered end, while a blocked end to

232 Stairs, Steps and Ramps

Figure 10.2c General view, Tulip staircase, Queen's House, Greenwich, 1629–35. (Inigo Jones)

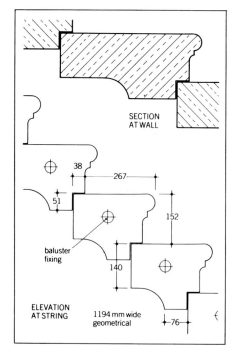

Figure 10.2d Detail of treads, tulip staircase

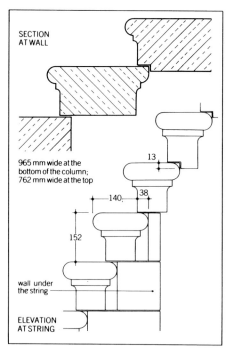

Figure 10.2e Details from cantilever steps, The Monument, City of London, circa 1671. (Sir Christopher Wren and Robert Hooke)

Stairs, Steps and Ramps 233

Figure 10.2f General view, The Great Stairs, Chatsworth House, Derbyshire, 1689–90. (William Talman, balustrade Tijou)

the nosing gives the necessary rebate at the tail of each tread. This detail is abandoned in the lower parts where a supporting wall exists towards the open well. A

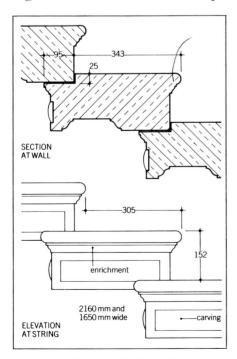

Figure 10.2g Detail of cantilever treads

further development in scale occurs with the geometrical stairs below the southwest Tower stairs at St Pauls Cathedral constructed by the master mason W. Kempster in (1706–8).

The most courageous cantilevers were made under Talman's direction at Chatsworth House, Derbyshire. The geometry follows a multiple turn staircase within a square shaft on the Palladio pattern illustrated in *Quattro Libri*. The cantilever dimension of 2 464 mm implies the cutting of vast blocks of sound local sandstone, the quarter landing slabs having sizes of around 2 600 mm^2! Each step weighed around 500 kg, the design above the first floor being reduced to more modest sizes of around 1 600 mm together with the introduction of carriage pieces. Talman was also involved with work at Hampton Court and it is thought likely that the modillion profile (Figure 10.2*b*) can be attributed to that source.

The tailing in of the tread ends is critical in terms of tightness of fit, the depth of bearing varying between 110 mm and 225 mm, the lesser dimension is common in nineteenth-century domestic work while Wren used 150 mm for the south-west Tower stairs at St Pauls and 225 mm at Hampton Court. In the writer's experience the crucial stabilizing role of the metal balustrade needs to be taken into account. Dismantling the balusters will often destabilize the treads. Such strengths may not be calculated but there is little doubt that stone cantilevers, and side mounted balusters, act together once assembled with leaded joints from metal to stone (refer back to Inigo Jones and Nicholas Stone details in Figures 10.2*c* and *d*).

An interesting new flight of cantilever stairs has been assembled at Da Costa

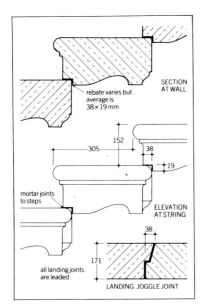

Figure 10.2h Modillion profile to stairs at Hampton Court, circa *1700*

House, Highgate by Russell Taylor as architect and Sam Price as engineer. Figures 10.2*i* and *j* reveal how the design has been adapted to the present day with artificial stone treads and mild steel balusters, the tailing in amounting to a half brick thickness; 1:1:6 mortar was used between the steps with stronger epoxy mortars for the dowels connecting the landing units. The designers had the useful experience of constructing a similar scale of staircase in traditional timber framing on the same job. The comparative costs between artificial stone and

Figure 10.2i General view, Da Costa House, Highgate, 1990. (Russell Taylor, engineer Sam Price, of Price and Myers)

timber were £22,000 and £28,000 at 1990 prices. This economic advantage shows that the tradition of Inigo Jones Tulip stairs may be revitalized and return to the architectural language of turret stairs. In conclusion appreciation is given for the considerable help from Russell Taylor and Sam Price in preparing this chapter and for their permission to use the illustrations.

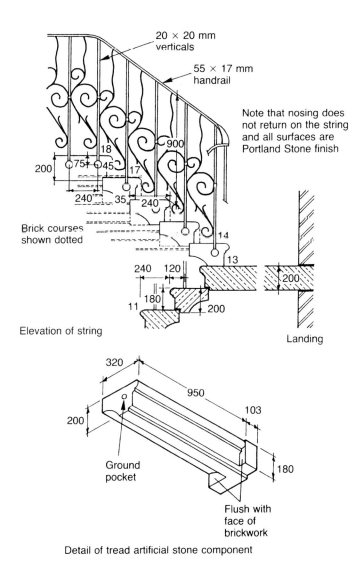

Figure 10.2j Detail of stair, Da Costa House

11 Design codes and procedures

To make sense of the British Building Regulations as they relate to stairs and means of escape is a form of purgatory. Thoughts turn to worthy colleagues who have departed from the pleasures of building into the domain of lawyers and law makers, and are now designing a paper world for others to construct. Such a world is depicted in the gauche illustrations that pertain to Building Regulation guides, many led by an obsession with the Building Inspector's 100 mm ball already mentioned in Chapter 8. The Scots are afflicted with their own perverse variations, it will suffice at this stage to comment on the Regulations as they apply to England and Wales including inner London (the area governed by the former London County Council up to 1964).

11.1 Key matters within the Building Regulations (including the 1989 and 1992 Amendments)

The approved document K represents an attempt to reorganize matters, these commence with standard definitions.[1] Key diagrams are given in Figures 11.1*a*, *b* and *c* that set down the general requirements for stairs. There are immediate danger signals for those who have fallen down steps. For example the stair that commences on the door line – what an unpleasant way of leaving a room! The second, and more serious error, is that of handrailing omitted for the last two steps – bad light or haste can precipitate a fall since the 'hand hold' and stair run is linked in the mind's eye. The regulations recognize this danger for escape ramps and stairs and demand a handrail for the full length. What a pity that the law makers cannot be more consistent.

The reforms made by the new revised Building Regulations which came into force on 1st June 1992 show some improvement by simplifying Section K which deals with the able bodied, whilst references to paragraphs ADM relate to facilities for the disabled. BS 5588 and Home Office requirements are drawn into the performance criteria under the titles given fully in References 4 & 5 and which will be extended in future.

Rise and going relationships have already been discussed in arriving at standard stairs for flats and commercial buildings in Chapter 3 Section 3.2 and Chapter 4 Section 1. The practical limits for the various building categories are given in Figures 11.1*d*, *e* and *f*. The pitch limitation should be noted (42 degrees for private stairs, 35 degrees for assembly and institutional buildings as well as common stairs). The Codes call for relationship of tread to twice the riser (T + 2R) to fall between 550 and 700 mm. The range of permitted ratios are those falling within the heavy line on the graphs.

It is most important to judge the likely future of buildings when selecting stair categories. Private stairs imply individual

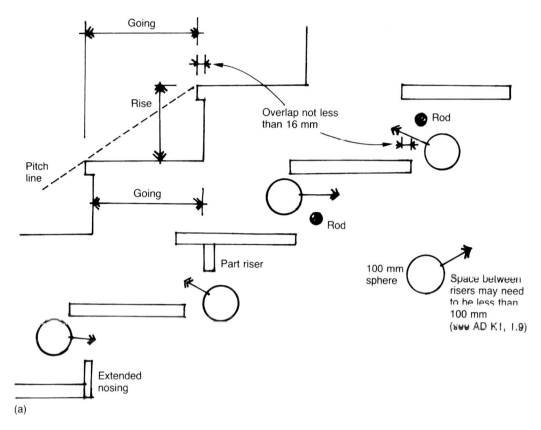

Figure 11.1 General requirements under Section K
a Rise and going

staircases within a dwelling, intended conversion to institutional use or into flats will mean upgrading standards and hence the selection of Figure 11.1*e* or *f* instead of 11.1*d* at the design stage.

Institutional and assembly buildings are premises usually occupied by young children, old people or the disabled. Stairs therefore have a shallower pitch of 35 degrees.

Stairs outside applications for Figure 11.1*d* and *e* are termed 'other stairs'. Commonly, these fall within the category of escape stairs in commercial buildings and blocks of flats. The same proviso exists regarding the pitch of 35 degrees.

There are other important rules that apply to all stairways.

- In any stairway there should not be more than 36 risers in consecutive flights, unless there is a change in the direction of at least 30 degrees.
- The ratio T + 2R equaling 500 mm minimum to 700 mm maximum has already been mentioned. There are permitted variations at tapered steps as Figure 11.1*g*.
- The rise of any step should generally be constant throughout its length and all steps in a flight should have the same rise and going. However, where a step adjoins the ground or paving outside a building they may be at a slope. In this case the rise of the step should be measured at the centre of the width of the flight.

238 Stairs, Steps and Ramps

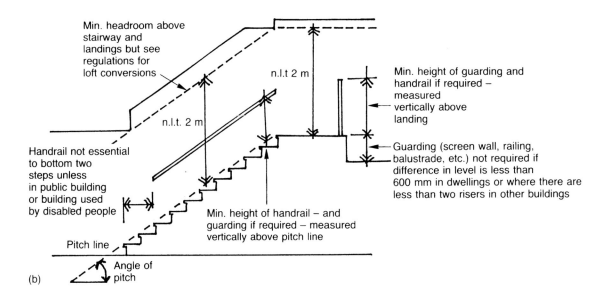

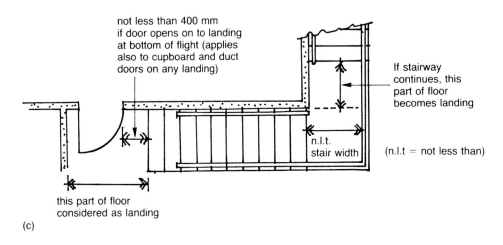

Figure 11.1b General requirements (section)
c General requirements (plan)

- Open risers are permitted in a stairway but for safety the treads should overlap each other by at least 15 mm.
- Each tread in a stairway should be level.

A basic guide concerning the domestic use of spiral stairs or winders is given in Chapter 3 Section 3.2.3, there are however specific rules for tapered treads as listed below and for more detail refer to Figure 11.1g.

- The minimum going at any part of a tread within the width of a stairway should not be less than 50 mm.
- The going should be measured:
  (i) at the centre point of the length or deemed length of a tread if the stairway is less than 1 m wide, and
  (ii) at points 270 mm from each end of the length or deemed length of a tread if a stairway is 1 m or more wide. (When referring to a set of consecutive tapered

Stairs, Steps and Ramps 239

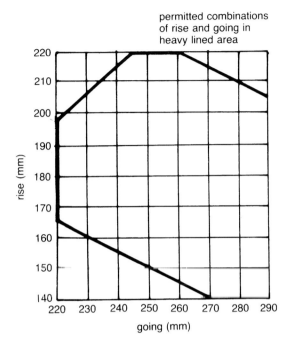

Figure 11.1d Practical limits for private stairway (rise and going)

treads of different lengths, the term 'deemed length' means the length of the shortest tread).
- All consecutive tapered treads in a flight should have the same taper.

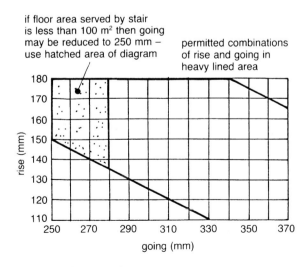

Figure 11.1e Practical limits (rise and going) – Institutional and Assembly Buildings

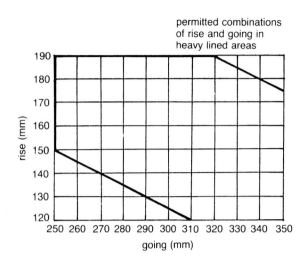

Figure 11.1f Practical limits (rise and going) – other buildings

Section ADK of the Regulations no longer has specific rules on the widths of stairs but one has to look to other sections such as ADB for fire safety and ADM for access and facilities for the disabled. Despite the new found freedom, the practical problems of handling furniture need to be considered and 800 mm is the minimum width. However, bathrooms or WCs can be accessed by stairs as narrow as 600 mm.

Sections ADB and ADM are restrictive on width and need to be checked against the building use (place of assembly, institutional use etc.) and the number of people using the premises. An extract of the key dimensions are given as follows.

- The narrowest permitted escape stair is 800 mm.
- Common stairs in flats have to be 1 000 mm but 1 100 mm if they are fire fighting stairs.
- A stair serving an assembly building has to be 1 100 mm unless the area served is smaller than 100 m^2.

240 Stairs, Steps and Ramps

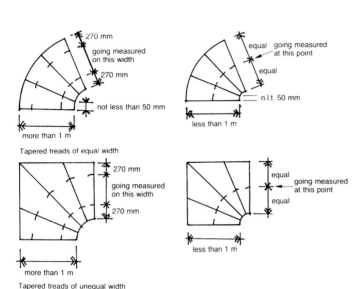

Figure 11.1g Tapered treads

Maximum number of persons	Minimum width (mm)
50	800[3]
110	900
220	1100
more than 220	5 per person

Figure 11.1h Widths of escape routes and exit

Maximum number of people in any storey	Stair width[1] (mm)
100	1000
120	1100
130	1200
140	1300
150	1400
160	1500
170	1600
180	1700
190	1800

Notes:
1. Stairs with a rise of more than 30 m should not be wider than 1400 mm unless provided with a central handrail (see para 4.6 [p.7.55]).
2. As an alternative to using this table, provided that the minimum width of a stair is at least 1000 mm, the width may be calculated from: $(P \times 10) - 100$ mm where P = the number of people on the most heavily occupied storey.

Figure 11.1i Minimum aggregate width of stairs designed for phased evacuation

- Stairs that are wider than 1 800 mm in an assembly building have to be divided by guard rails.
- Stair widths are measured at 1500 mm height.
- The widths depend upon the number of persons and are determined from Table 5 under Section AD B1 (refer to Figure 11.1h, j)
- The Regulations assume that in buildings with two escape staircases one will be disabled by fire. The widths given for *each* stair therefore have to be the 'Table 5' size since either may have to take the full traffic load in emergencies.
- Long flights over 30 m in descent to be 1 400 mm maximum width, which implies a total traffic limit of 700 persons per stair (using Figure 11.1j).

- Stairs over 1 400 mm wide may have to jump to 1 800 mm wide to comply with the central guard rail rule. Typical traffic numbers are given in the extract from Table 5 of AD B1. This table is also the guide in designing stairs for total evacuation of open planned buildings, basements, recreational and residential buildings (Figure 11.1h)
- Phased evacuation permits narrower stairs since safe collection points have to be provided at various floor locations in the height of the building. Table 8 of AD B1 contains the key data (Figure 11.1i). Buildings planned for phased evacuation require specific protection, namely that each floor is a compartment floor, stairs always relate to protected lobby or passage approach, a fire warning system and intercom are

Stairs, Steps and Ramps 241

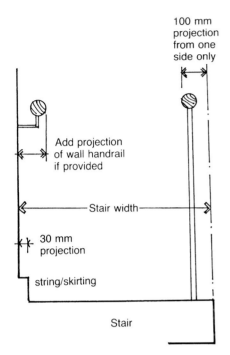

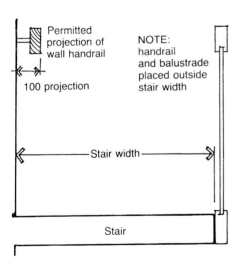

Figure 11.1j Handrail projection into stair widths

needed between fire wardens (floor by floor) and the control point. Sprinklers are required throughout buildings which exceed 30 metres in height.
- Section ADB permits strings and handrails to project into the stair width. It would be wise to establish with the Fire Brigade or Licensing Authority that the 100 mm projection will be accepted.

The regulations and codes in the UK are very specific about forms of lifts and stairs that are *not* acceptable as a means of escape in case of fire.

- Lifts unless evacuation lifts.
- Portable or throw away ladders.
- Fold down ladders.
- Escalators (though travelators are accepted if switched off in case of fire).
- Accommodation stairs (except for small shops with a total area of all floors less than 90 m^2 and limited to basement, ground and first floor level).

The regulations concerning construction are vague since matters are referred to in Codes of Practice or British Standards as outlined in section 11.5. Section AD B2/3/4 (sections 1 to 6) does contain guidance for exceptions to the rules. The intention is that staircases and landings in all classes of building are constructed of limited combustibility (though not exactly defined) unless the stairs are located as follows:

- In a maisonette.
- In a building or compartment for which the elements of structure require a fire resistance period of less than one hour.
- In a ground or upper storey (but not a basement) of a building comprising flats or mainsonettes (but not more than three storeys).
- An external stair between the ground and a floor or roof which is not more than 6 m above ground.
- A stair that is not within a protected shaft in a building or compartment which is used as a shop (called accommodation stairs in Chapter 4).

It is wise to check the interpretation of these points with the Fire Authority.

11.2 Ramps and guarding of stairs, etc.

The categories for the use of ramps are similar to those set down for stairways, the width not being less than that required for the same category of stair.

- The slope of any ramp should not be more than 1 in 12;
- The length of any ramp in a stepped ramp should be between one and two metres, the intermediate steps being level.

The last proviso makes a nonsense of the notion of wheelchair access.

The guarding rules are contained in Figure 11.2a for ramps and stairs and one can see at a glance the dichotomy that faces designers trying to achieve alignment of balustrades to flights and landings. The situation regarding glazing is that the material has to be laminated safety glass, toughened glass or glass blocks, acrylic is not mentioned and will clearly be the subject of a waiver. The limitations imposed by the gap (in open stairs and balustrades), which is not to be passed by the 100 mm ball are specific to dwellings, institutional buildings (where children under five will be present) and any other residential building. The 1989 regulations do not permit balustrades that can be easily climbed by children, where they might be at risk.

It is easy to imagine the situation under the British legal system that such restrictions are difficult to avoid with any common or public stairs where children will be present; for example leisure centres, hotels or museums. The designer needs to be wary in terms of insurance risks. It is often easier within the private domain of a house where the onus for compliance and insurance can be placed on the owner. Many prefer to have no balustrade at all while others keep a gate at the top and the bottom of their stairs as a guard against children entering a danger zone.

If one perseveres with common sense, then clearly a vertical balustrade, mesh, sheet or solid material will provide the safest protection for small children. A further application of common sense will reveal that solid walled balconies or landings will tempt children to bring chairs or similar items into place in order to sit astride the balcony top, not to mention sliding down the solid balustrade! On second thoughts it is a matter of rods, mesh or sheet in terms of safety for inventive or uncontrolled children, whether small or large.

Vehicle ramps or areas of buildings accessible by vehicles are required to have barriers in order to protect people. Document K2/3 in the Regulations provides guidance on barriers and is shown in Figure 11.2b.[2]

11.3 Means of escape in case of fire (general matters under The Building Act of 1984)

The national Building Regulations make general provisions on access and exit matters and these can be summarized as follows.

AD K2/3, Section 3

Diagram 11 Guarding design

Building category and location		Strength	Height (h)	
Single family dwellings	stairs, landings, ramps, edges of internal floors	0.36 kN/m	900 mm for all elements	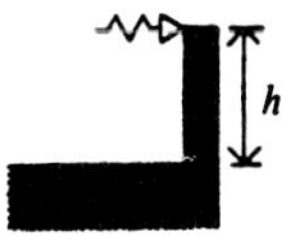
	external balconies and edges of roof	0.74 kN/m	1100 mm	
Factories & warehouses (light traffic)	stairs, ramps	0.36 kN/m	900 mm	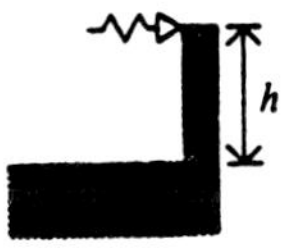
	landings and edges of floor	0.36 kN/m	1100 mm	
Residential, institutional, educational, office, and public buildings	all locations	0.74 kN/m	900 mm for flights otherwise 1100 mm	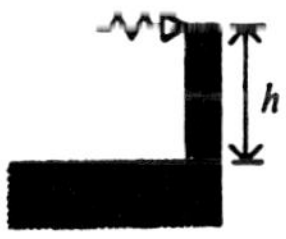
Assembly	530 mm in front of fixed seating	refer to BS 6399: part 1	800 mm (h_1)	
	all other locations	refer to BS 6399: part 1	900 mm for flights elsewhere 1100 mm (h_2)	
Retail	all locations	1.5 kN/m	900 mm for flights otherwise 1100 mm	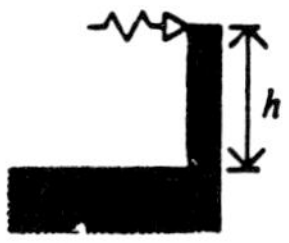
All buildings except roof windows in loft extensions (see AD B1 and Fig. 7.13(b))	at opening windows		800 mm	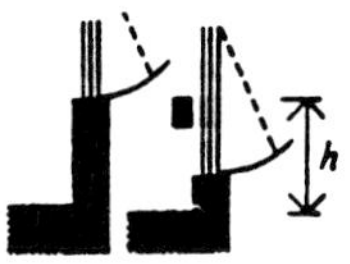
	at glazing to changes of level	to provide containment	below 800 mm	

Figure 11.2a Guarding of stairs and ramps

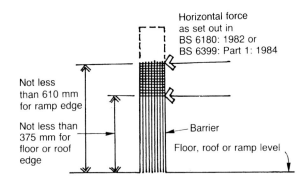

Figure 11.2b Vehicle Barriers (ramps, etc.)

- *1984 Act Section 24: access and exit* This section governs access and exit in public buildings, and places of public assembly as well as those licensed for entertainment, music and dance. The Fire Authority has to agree the details; disputes can be determined by the Magistrates Court.
- *1984 Act Section 6 and 7: deposit of plans* Section 6 and 7 requires deposit of plans that show the intended arrangement and stipulate that plans shall be updated to show what has been finally built and agreed.
- *Local legislation* Many local authorities have created special powers relevant to building control and these should be checked. Under these local provisions there often exist extra safety requirements for basement garages e.g. the lobby approach to stairs. Another common requirement is that buildings in excess of 18.3 m height have fire alarm systems and that adequate access exists for the fire brigade to reach lifts and stairs.
- *1984 Act: means of escape above 6 or 4.5 m* This provision covers escape stairs where there is a storey over 6 or 4.5 m above ground in specific buildings namely hotels, boarding houses and hospitals.

11.4 Buildings of excess height and excess cubical content

Such buildings are still governed within inner London by Section 20 of the London Building (Amendment) Act 1939. The provisos affect buildings higher than 30 m or 25 m (if the area of the building exceeds 930 m^2) or for large trade buildings over 7 100 m^3. The means of escape in case of fire under Section 20 is still controlled by sections 33 and 34 of the 1939 Act. The requirements are more stringent than those in Code of Practice 3 clause IV or in BS 5588.

11.5 Mandatory rules for means of escape in case of fire

The mandatory rules on the means of escape in case of a fire are prescribed in Part B of Schedule 1 to the Building Regulations (1991), known as 'Approved Document B1'. A prime requirement is the provision of adequate exits and protected escape routes. The other salient factor is the fire resisting standard of materials in such zones as required by the Fire Authority. Compartment floors where pierced by stair wells or lift shafts have to be enclosed within a fire-resisting structure termed protected shaft, with openings formed by fire resisting doors or shutters.

The general thrust of the 1991 Regulations can be summarized in order of priority for the construction of buildings.

DESIGN	DETAILING
• To allow the occupants to safely escape.	Properly placed exits and protected exit routes.
• For the building to resist collapse until the occupants are clear and to prevent further spread of fire across compartment zones.	Achieving fire resistance of structure (floors, walls, roofs and frames).
• To keep the fire from spreading within the building or to another building.	Provision of compartment zones, setting standards for existing walls, controlling flame spread across wall and ceiling surfaces, sealing concealed space or ducts. Setting standards of fire resistance, also limiting flame spread for roofing.
• To give access for firefighters.	Providing protected passages and shafts.

These priorities differ from those required by insurers and it is important to check with the building owner and their insurers on protective measures to save the building and its contents, vis à vis the installation of sprinklers and the further curtailment of compartment zones. The complexities of the mandatory rules are best explained by placing them in 'design order'. This gives a step by step process through the decisions to be made concerning means of escape.

First stage
Deciding upon the building use, termed 'classification of purpose groups' in AD 12, Appendix D. Table D 1.

Second stage
Establishing the number of people using the building. Such figures may be stated in the programme (e.g. theatre seating capacity) or dependent upon areas created and then computation for occupancy rates. The floor space factors are given in AD B1, Table 1 (see Figure 11.3). If escape by people in wheelchairs has to be considered the designer should refer to paragraph 5.36 in BS 5588 Part 8[3].

Areas of special fire risk should be considered such as escape from boiler rooms and fuel stores or places involved in manufacture.

Mixed use buildings normally require separate escape measures for each category of use, e.g. residential versus assembly versus institutional.

Third stage
Designing horizontal routes for escape within the permitted travel distances and establishing the required widths. The widths will have to be maintained for stairs to be negotiated in stage four. Widths at doorways to stairs are measured between jambs, excluding the door leaves.

Figure 11.4 is a key diagram for various categories of use. The designer needs to study both the approved document B1 on means of escape and BS 5588 (parts 1 to 10)[4], the latter takes precedent over document B1 for offices and large stores and enclosures. Perseverance is needed to unravel the

Type of accommodation(1)(6)	Floor space factor m²/person
1. Standing spectator areas	0.3
2. Amusement arcade, Assembly hall (including a general purpose place of assembly), Bar (including a lounge bar), Bingo hall, Dance floor or hall, Club, Crush hall, Venue for pop concert and similar events, Queing area	0.5
3. Concourse or shopping mall (2)	0.75
4. Committee room, Common room, Conference room, Dining room, Licensed betting office (public area), Lounge (other than a lounge bar), Meeting room, Reading room, Restaurant, Staff room, Waiting room (3)	1.0
5. Exhibition hall	1.5
6. Shop sales area (4), Skating rink	2.0
7. Art gallery, Dormitory, Factory production area, Office (open-plan exceeding 60 m²), Workshop	5.0
8. Kitchen, Library, Office (other than in 7 above), Shop sales area (5)	7.0
9. Bedroom or Study-bedroom	8.0
10. Bed-sitting room, Billiards room	10.0
11. Storage and warehousing	30.0
12. Car park	two persons per parking space

Notes:
1. Where accommodation is not directly covered by the descriptions given, a reasonable value based on a similar use may be selected.
2. Refer to section 4 of BS 5588: Part 10 for detailed guidance on the calculation of occupancy in common public areas in shopping complexes.
3. Alternatively the occupant capacity may be taken as the number of fixed seats provided, if the occupants will normally be seated.
4. Shops excluding those under item 8, but including supermarkets and department stores (all sales areas), shops for personal services such as hairdressing and shops for the delivery or collection of goods for cleaning, repair or other treatment or for members of the public themselves carrying out such cleaning, repair or other treatment.
5. Shops (excluding those in covered shopping complexes, and excluding department stores) trading predominantly in furniture, floor coverings, cycles, prams, large domestic appliances or other bulky goods, or trading on a wholesale self-selection basis (cash and carry).
6. If there is to be mixed use, the most onerous factor(s) should be applied.

Figure 11.3 Floor space factors

Gordian knot of rules since other sources have to be studied such as the Home Office and sundry ministerial publications[5]. Inner London still has specific reference to escape in case of fire from large building volumes (see Section 4 and Reference 6 for data).

Fourth stage
Designing vertical routes for escape, namely stairs for the able bodied, evacuation lifts for the disabled and access shafts (stairs and lifts) for fire fighters. The same references occur as for stage three. Settling the number and width of stairs

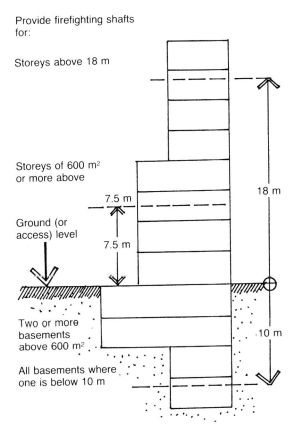

Figure 11.4 Fire fighting shafts

involved depends upon the horizontal travel distance to stairs. Stairs without lobby approach may have to be discounted. Design for evacuation falls into two categories, total evacuation for basements, open plans, residential and assembly building. Partial or phased evacuation may apply for institutional buildings and office premises with safe lobbies or zones forming collection points for rescue. For example flat roofs, balconies or open sided lobbies.

Fifth stage

Agreeing constructional standards for stair and lift enclosures as well as protected lobbies and passages. Generally, this implies 30 minutes fire resisting enclosure. There are however exceptions. Doors, glazing and ironmongery have to be specified to accord with AD B1 and in relation to paragraphs 5.12 (doors) and 5.5–5.9 (glass). Ironmongery patterns are best agreed with insurer's requirements[7]. The relevant British Standard is BS 5588 part 1, 1990.

11.6 Other legislative requirements

In Britain the Acts related to Fire and Buildings are cumulative. This no doubt reflects the fact that government lawyers are paid better for their pains than architects, hence the duplication of work rather than a comprehensive effort to put fire regulations into a single document. A review of the diverse Acts follows:

- *The Fire Precautions Act 1971* A fire certificate is required for designated premises.
  (a) *Hotels and boarding houses* except those where sleeping accommodation is provided for not more than six people and none is provided above first floor or below ground floor.
  (b) *Factories, offices and railway premises* where more than twenty people are employed other than on the ground floor. Also where explosive or highly flammable materials are stored or used.
  (c) *Institutions for treatment or care.*

(d) Buildings or places for entertainment, recreation, instruction or used by clubs, societies or associations.

(e) Buildings for teaching, training or research (places for public worship are exempt).

(f) Public access whether paid access – or otherwise, (single dwelling houses are exempt).

Legal pens are always busy and further revisions were made under the appropriately named Fire Precaution, (Modifications) Regulations 1976.

- *The Housing Act 1980 & 1985* Means of escape required in houses of multiple occupancy and where the property has three or more storeys and a combined floor area exceeding 500 m^2. However if a *new* dwelling complies with AD B1 then this is acceptable for multiple occupation.
- *Health & Safety at Work Act 1974* This applies to specialized industrial and storage premises and which have separate certification.

11.7 1992 Revisions

Figure 11.4 features the key changes made in the 1992 version of the National Building Regulations, the points are as follows and drawn from the latest edition of the 'Easiregs'[8] prepared by Henry Haverstock for *Building Design*.

11.7.1 Provision of fire fighting shafts (under B5 (1) & (2))

Access to properties in order to fight fires are familiar topics discussed with Fire Brigades. These provisions are now incorporated as from June 1992, Figure 11.4 gives the salient data.

High buildings and those with deep or extensive basements need fire fighting shafts located in such a way that every part of each floor is within 60 m of a shaft entrance, or 40 m where the floor layout is not known. The shaft or shafts need fire fighting facilities, firemans lift, stairs, lobby with dry/wet riser outlet and intercom. The number of such shafts are negotiable but the interconnection to the street entrance should be short (maximum passage of 18 m). Wet risers are insisted upon where a building floor exists more than 60 m from ground or access level. The Code cross references to BS 5306 parts 0, 1, 2 and 3, BS 5588 parts 5 and 10, BS 5839 part 1.

11.7.2 Brigade access to buildings (under B5 (2))

Detailed requirements for access around the site via roads and hardstandings are also scheduled and graded according to the area of the building. The significant point relates to buildings with built-in fire mains where Brigade access has to be within 18 m of a position within sight of the hose inlet.

11.7.3 Escape from houses (under B1)

The most pertinent aspect relates to the protected stair shaft needed in three storey houses up to the second floor and that smoke alarms have to be fitted in stair halls (whether two or three storeys)

11.7.4 Escape from flats (under B1)

Ground and first floor flats (including ground floor maisonettes) are treated as houses, namely that the occupants can jump from the windows. The general provision of smoke alarms is made to all units. Flats and maisonettes at second floor and above are subject to mandatory rules for escape which are summarized in Figures 11.5a, b and c. Protected hallways within flats are limited to 9 m in length, the same dimension applies to the escape routes in small flats, such as bed-sitting or one bedroom units, if the kitchen is sited away from the flat entrance. Flats planned more spaciously will need alternative means of escape to balconies and secondary stairs. The same applies to all maisonettes at second floor and above. Such secondary stairs can be external but are

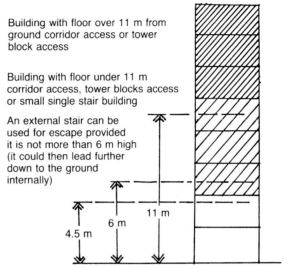

Figure 11.5 Escape from flats
a Heights and means of escape

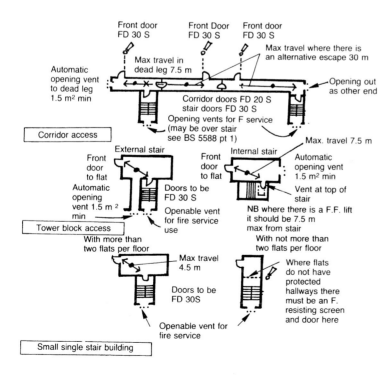

Figure 11.5b Corridors and lobbies (escape from flats)

250 Stairs, Steps and Ramps

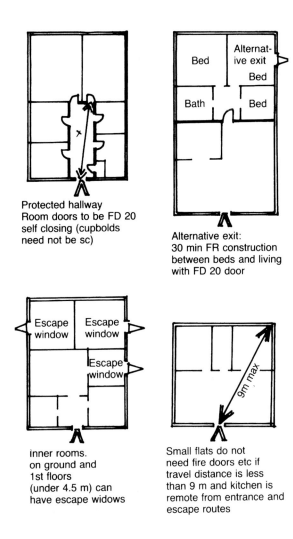

Figure 11.5c Flat plans (escape from flats)

distances (open plans). See Figures 11.6*a*, *b* and *c* and which refer to all levels (ground and upwards). Clearly the number of exits from a particular floor is also dictated by the occupancy, see table in Figure 11.6*d*.

Vertical escape stairs have to be duplicated except in the following circumstances.

- Buildings of three or less storeys.
- For building heights not greater than 11 m (this is measured from the floor plane of the top floor to the lowest ground level adjacent to the building)
- Where travel distances or numbers warrant more than one stair.

A typical vertical core is shown in Figure 11.6*e* for an escape stair, lobby and fire fighting lift with external fenestration to stair and lobby. Such lobbies cannot be used as through routes to other stairs unless sub-divided by fire doors. See Figure 11.4 for fire fighting shaft provision.

limited to 6 m unless protected from snow and ice. Primary escape stairs should terminate at street level in order that basement stairs are separate with a ventilated lobby approached from the lower level to give smoke checks to the ground floor hall.

11.7.5 Escape from offices (BS 5588 Part 3, Codes of Practice for Office Buildings[4])

The horizontal escape is limited by travel distances (passage and room) or direct

11.8 Procedure to obtain a Fire Certificate (Fire Precaution Act 1971)

The requirement to deposit plans calls for careful procedure. A well tried method favoured by fire officers is a set of drawings, plans and sections (with floor levels and uses noted) which relate only to means of escape provisions, with details shown of lifts, stairs, protected passages and shafts as well as Fire Brigade access within the site. Those details do not need to contain contractors' information but simply to record to scale the general arrangement with a code for fire doors

Stairs, Steps and Ramps 251

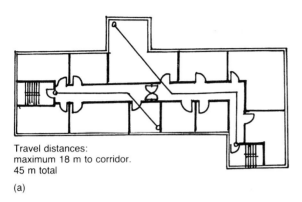

Travel distances:
maximum 18 m to corridor.
45 m total

(a)

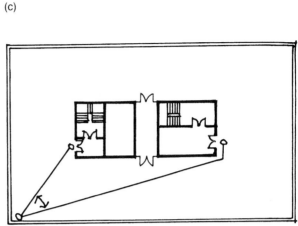

Direct distances – central core
maximum 12 m if angle is less than 45°

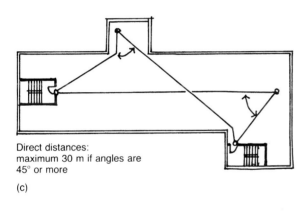

Direct distances:
maximum 30 m if angles are
45° or more

(c)

Maximum number of persons	Minimum number of escape routes/exits
500	2
1000	3
2000	4
4000	5
7000	6
11000	7
16000	8
more than 16000	8

(d) Table for number of escape routes

(e) Vertical core layout

Figure 11.6 Escape from offices

and fire resisting enclosures and screens. Stair directions and tread numbers should be stated as well as girths of stairs and landings with tread-to-rise relationship. Data such as the actual fire rating and agreed construction can be formulated in a typed report. The eventual licence can incorporate this information and can be easily updated if changes in layout or specification occur. It also provides both client and fire officer with documents that pertain to the licence without confusion caused by extraneous constructional information. Another version of the negatives can be prepared for noting fire alarms, signs and emergency lighting.

In conversion work, Fire Authorities often ask for plans of the rest of the building and which can incur general up-grading of the fabric. Subsequently, Fire Authorities are entitled to make inspections to satisfy that the 'means of escape' are in effect and safe and that fire fighting equipment and warning signs and notices are in place. A copy of the Fire Certificate has to be kept on the premises. Lack of inspection by the Authority does not invalidate the Certificate. The Fire Certificate assumes that the building is properly managed, failure to manage could mean prosecution and prohibition in use.

11.9 New technology

Today's technology for fire protection includes smoke detectors, linked to automatic door releases and which overcome the annoyance of lobby approach to stairs in high risk locations. Today the stair enclosure doors can operate on a self-closing device whilst the second set of doors can be released by electromagnets.

A further change is the acceptance of positive air pressure within the escape routes and stair halls to expel smoke that may enter from adjacent accommodation. This factor simplifies escape provisions – refer to the Economist Building featured in Figure 4.14 where the 1964 plans (under the former LCC regulations) are compared with the 1991 version taking into account the latest technology. The Loss Prevention Council[9] is a useful authority on up to date procedures. They publish a number of guides and advise on fire fighting facilities. British Standards to refer to are BS 5588 Part 1, 1990 (Point type smoke detectors) and Part 4 1978 (Code of practice for smoke control in protected escape routes using pressurization).

11.10 American codes

The final leg of this chapter (Sections 11.11 to 11.16) has been prepared by Ian Smith ARIBA AIA, a British architect who has been working in New York for more than 30 years. He reviews two codes that apply in the eastern and western parts of the USA and which have the advantage of brevity.

11.11 A review of code requirements governing stairs and ramps in the USA

The USA has a population of over 250 million, has a land mass twice the size of Europe and is divided into 50 separate states, each of whom jealously guard their right to establish their own laws. It can therefore be hardly surprising that, throughout the country, a great number of different building codes are in force.

The larger cities all impose their own codes which, being written by lawyers, are provided with enough built-in contradictions to keep their framers in business. The New York City code, for example, is so opaque and so resistant to rational interpretation that a new type of professional has been created to deal with it; part architect, part engineer, part lawyer known as an 'Expediter'. For a hefty fee the expediter shepherds the architect's drawings through the arcane workings of the borough building departments, surfacing from time to time with necessary pieces of documentation; emerging finally with the certificate of occupancy in hand.

Most smaller towns and cities, however, have found it expedient to adopt one of the two National Codes that have now been widely used for many years, both of which provide a consulting and inspection service. These codes are:

- The Uniform Building Code
  First enacted by the International Conference of Building Officials in Phoenix, Arizona in 1927. Their present headquarters are in Whittier, California.
- The BOCA National Building Code
  Published by the Building Officials and Code Administrators International, Inc. (BOCA), first enacted in 1950. Their present headquarters are in Chicago, Illinois with regional offices in Columbus, Ohio, Tulsa, Oklahoma and Trevose, Pennsylvania.

In very broad terms it can be said that the Uniform Building Code is favoured west of the Mississippi River, the BOCA National Building Code to the east.

For the purpose of this review the BOCA code will be discussed, with comments on the UBC to those areas where there is a significant divergence from BOCA. The summaries are based largely upon stair requirements for commercial buildings.

11.12 Planning principles governing the number and location of exit stairways

The following review is restricted to 'required' stairs; that is to say stairs used exclusively for the evacuation of a floor to the open air at ground level, contained within a fire-rated enclosure.

The location and number of legally required exit stairs from any floor is determined by the total occupancy of the floor, the distance of travel from any point on the floor to the nearest stair, and the basic principle that stairs should be remote one from another.

The occupancy of the floor is calculated by referring to a table that sets out the maximum floor area allowance per occupant for a wide range of uses. The allowance for a usage not specifically included in the table may be determined by negotiation with the building official.

Once the occupancy has been determined the minimum number of stairways is calculated as in Table 11.1.

Table 11.1

Occupancy load	Minimum number of stairs
500 or less	2
501 – 1000	3
Over 1000	4

The second factor determining the number of stairs required is the limitation of the distance of travel (measured as an exact *path of travel*) from any point on the floor nearest stair or fire-rated corridor leading to a stair. This distance varies in accordance to usage and whether or not the space is served by sprinklers. For example, in such general usages as assembly, business, mercantile or educational the maximum path of travel distance is 250 ft (76 m) if the space has sprinklers and 200 ft (61 m) if it has not. Hospitals are more restricted; warehouses less. The final planning restraint is the doctrine of 'remoteness' which governs the location of stairs in relation to one another. The rule states that where only two stairs are required they should be no closer (when measured in a straight line) than half of a straight line connecting the two most remote corners of the space. The distance of separation may be halved if the space is sprinklered; a concession, it should be noted, that is not allowed in the Uniform Building Code.

In large floors where more than two stairs are required the rule governing path of travel ensures that stairs are well separated, but in any case the code inspector will insist on a reasonable distribution of stairs.

11.12.1 Width of stairs

Having calculated the number of stairs it is now necessary to calculate the total stair width. This is, of course, directly related to the occupancy of the floor. The method of calculation, however, varies radically between BOCA and UBC and I will therefore review both.

11.12.2 Stair width calculation, BOCA Code

The total stair width is calculated by using a *unit width* multiplication by the number of occupants. The unit of width varies from use to use and is significantly less when the space is sprinklered. For example:

a For general use such as assembly, business, educational, mercantile, residential, storage or manufacture the unit width is 0.2 in (5 mm) if the space is sprinklered, and 0.3 in (7.5 mm) if not.
b In hospitals the unit width is 0.3 in (7.5 mm) if sprinklered and 1 in (25 mm) if not.

Exit stair width is calculated for the *occupancy of the floor only*, that is to say, the stair 'load' *is not cumulative*.

11.12.3 Stair width calculations using UBC

Total stair width is calculated by dividing the total occupancy by 50, expressing the result in feet. No distinction is made between use groups and no concessions are granted for sprinklered floors.

The major difference between BOCA and UBC however is that UBC calculates the stair loading in a multi-storey building on a *cumulative basis*. The rule is that the stair must accommodate 50 per cent of the load of the floor immediately above plus 25 per cent of the load of the floor above that.

It is therefore immediately clear that in a building of more than two floors UBC is a much stricter code than BOCA. As a practical matter however the different ways of calculating stair width do not become a design issue until the floor occupancy reaches a little over 200; as for example, in a multi-storey office building with a uniform single floor area of 20,000 ft² (1 860 m²). In a floor of this size BOCA requires a total stair width of 60 in (1.524 m) while UBC (bearing in mind that the load is cumulative) requires a width of 84 in (2.134 m). Both are therefore still below the mandated minimum requirement of two stairs of 44 in width each, i.e. 88 in (2.235 m).

The UBC requirements do become onerous, however, in buildings such as department stores where an upper floor occupancy is often as high as 1 500 persons.

11.13 Detailed design of stairways

a *Width*
No stair may be less than 44 in (1.118 m) wide. All stairs from a floor should be approximately the same width to achieve an even distribution of existing load.

b *Landings*
The landings must be the same width as the stairs, except that landings connecting flights in a straight run need not be more than 48 in (1.219 m).

c *Headroom*
Clear headroom measured vertically from the nosing of a stair or floor landing must not be less than 6 ft 8 in (2.032 m).

d *Vertical rise*
Maximum vertical rise between landings or intermediate platforms: 12 ft 0 in (3.658 m).

e *Treads and risers*
There are no regulations regarding pitch.
Maximum riser: 7 in (178 mm)
Minimum riser: 4 in (102 mm)
Minimum going: 11 in (279 mm)
No winders allowed.

f *Handrails*[4]
Handrails should present a continuous gripping surface without obstructions such as newels. There must be a clear $1\frac{1}{2}$ in (38 mm) from rail to wall and the rail may project into the stair a maximum of $3\frac{1}{2}$ in (39 mm). The rail must be no more than 38 in (9 655 mm) above the stair nosing or landing floor and no less than 34 in (864 mm).

Handrail ends must continue horizontally at least 12 in (305 mm) beyond the top nosing of a flight. At the bottom of a flight the handrail must continue at the same slope for the equivalent of one tread beyond the bottom nosing and then continue horizontally for a minimum of 12 in (305 mm).

Intermediate handrails are required so that no part of a stair is more than 30 in (762 mm) from a handrail.

g *Guardrails*
In educational buildings the guardrails at stairs and landings should not be less than 42 in (1.067 m). In all other use groups minimum height: 34 in (864 mm). Open guards (i.e. open balustrades) should have balusters or other constructions that will not pass a 4 in (102 mm) sphere.

11.14 General notes

The foregoing review has covered the rules governing the location, size and detail of the stairs themselves. Very specific rules also govern the sizes of the corridors or aisles leading to the stairs, the doors that open into the stairways and the doors that exit from them.

11.14.1 Aisles and corridors

Aisles (as in a department store for example) or corridors leading to any stair must be at least wide enough to match the capacity of the stair it serves.

11.14.2 Doors

Stair access doors must open in the direction of travel, i.e., into the stair enclosure. When doors are fully open they may intrude into the landing space no more than 7 in (178 mm). Doors must be self closing, be unlocked but equipped with a latch that will open automatically under pressure, and have a fire rating matching that of the enclosure. Doors exiting from stairs must be equipped with panic hardware. All doors, in or out, must be a minimum width of 32 in (813 mm) and match the capacity of the stair.

11.14.3 Stair enclosures

All stairs required as a legal means of egress must be constructed of non-combustible materials and must be within a fire rated enclosure. For buildings of more than four floors the rating is generally 2 hours; 1 hour if less. A 2 hour fire rating can be achieved with an assembly comprising 6 in (152 mm) galvanized metal studs with two layers of 5/8 in (16 mm) 'fire code' gypsum board (equal to Fireline board) on each side.

11.14.4 Lighting

All stairs must be equipped with an emergency lighting system. In smaller buildings the system may be energized by battery packs; in large buildings such as hospitals and department stores a generator set that kicks in immediately on the failure of electrical service, is required.

11.15 Domestic stairs

Stairs inside houses or apartments are, of course, much less restricted than legally required exit stairs. They need not be enclosed, risers may be a maximum of $8\frac{1}{4}$ in (210 mm) and the 'going' a minimum of 9 in (229 mm). Spiral stairs and winders are also permitted with some dimensional restrictions.

11.16 Ramps

The Americans with Disabilities Act of 1990 became effective on 26 January 1992. Its provisions, in the form of guidelines for the design of new buildings and for retro-fitting old ones, are not yet part of any building code. However, non-compliance with the Act's provision leaves the building owner open to suit on the grounds of discrimination on the basis of disability. The owner of an existing building is obliged to do everything possible to comply with ADA but it is recognized that some of the Act's requirements are not achievable. In new construction, however, no such excuse is possible and non-compliance opens up the very real possibility of a suit being brought against all those associated with a project: owner, architect and contractors.

ADA is a Federal Act and is applicable in every state. In dealing with this section therefore, UBC and BOCA have been put aside for a review of ADA rules and regulations.

a *Slope and rise*
   Maximum slope shall be: 1:12
   Maximum rise between landings shall be: 30 in (760 mm)
b *Clear width*
   Minimum clear width shall be 36 in (915 mm)
c *Landings*
   Level landings shall be provided at the top and bottom of each ramp run. Landings shall have a length of 60 in (1.525 m) clear and if the ramp changes direction at the landing, the minimum size shall be 60 in × 60 in (1.525 m × 1.525 m)
   Doorways at landings shall have a flat area in front of the door (in its closed position) of 60 in (1.525 m).
d *Handrails*
   If the ramp is over 6 in handrails must be provided on both sides. If handrails are not continuous they shall extend 12 in horizontally beyond the end of the ramp. Handrails shall be 34 in–38 in (865 mm–965 mm) above the ramp landing.
e *Edge protection*
   Ramps and landings with drop-offs shall have a minimum 2 in (50 mm) curb below in addition to standards supporting the handrails to prevent wheelchairs from sliding off the ramp.

References

[1] Standard definitions under the National Building Regulations (1989) in the UK:
   Circulation space This includes protected stairways as well as means of access between rooms and exit.
   Common stairway Stairways which serve two or more dwellings.
   Evacuation lift Lift used to evacuate disabled people in case of fire.
   Final exit Termination of escape to a safe zone . . . street, open walkway or open space.
   Going the distance measured in plan across the tread less any overlap with the next tread above or below.
   Pitch may be defined as the angle between the pitch line and the horizontal.
   Pitch line may be defined as a notional line connecting the nosings of all treads in a flight. The line is taken so as to form the greatest possible angle with the horizontal, subject to the special requirements for tapered steps.
   Platform floor Deemed to be a floor over a concealed space intended to have services.
   Pressurisation Mechanical technique to prevent ingress of smoke to protected egress route.
   Private stairway Stairways in or serving two or more dwellings.
   Rise the vertical distance between the top surfaces of two consecutive treads.
   Stairway Flights and landings included in this term.
[2] For greater detail on vehicle barriers refer to BS 6189 (1982) Code of Practice for Protective Barriers in and about Buildings. These cover circumstances where vehicles weigh more than 2.5 tonnes or are moving at greater speeds than 16 km/hour.
[3] BS 5588 Part 8 Code of Practice for Means of Escape for disabled people. For remainder see footnote 4.
[4] Rest of BS 5599 Fire precautions in the design, construction and use of buildings.
   Part 1 Residential buildings.
   Part 2 Shops.
   Part 3 Office buildings.
   Part 4 Smoke control.
   Part 5 Fire fighting stairways and lifts.
   Part 6 Assembly buildings.
   Part 7 Atrium buildings (draft).
   Part 8 See note 3 above.
   Part 9 Ventilation and air conditioning ductwork.
   Part 10 Enclosed shopping complexes.
[6] Home office and other Ministerial Publications.
   * Home Office Guide to fire precautions in existing places of entertainment. Home Office 1990.

- * DHSS Firecode. Health and Social Security. HMSO 1987
- * DES Building bulletin 7. Fire and design of educational buildings. HMSO 1988
- * Home Office Guide to safety at sports grounds. Home Office/Scottish Office HMSO 1990

[6] Section 20 Building and related matters. The source of documentation is LDSA Publications, PO Box 15, London, SW6 3TU, listed as follows.
Fire Safety Guides
1. Section 20 Buildings
2. Atrium Buildings.
3. Phased evacuation from office buildings

[7] Insurers Requirements can be answered in principle by the Association of British Insurers (Aldermary House, Queen Street, London EC4 1NT). Specific advice upon patterns of fittings is given by the Guild of Architectural Ironmongers (8 Stepney Green, London, E1 3JU) for ironmongery and types of doors that form escape stair enclosures or lobbies.

[8] 'Easiregs'. The latest collection of this valuable guidance is titled 'Easibrief' and was published for Building Design by Morgan-Grampian (Construction Press) Ltd in 1993. Many of the figures in this chapter have been redrawn from this source, courtesy of Henry Haverstock.

[9] The Loss Prevention Council, 140 Aldersgate Street, London EC1A 4HY.

12 Elevators and mechanical circulation

12.1 Introduction

The topic of machines for moving people is a manual in itself. The mechanical means whether elevator – lift, escalator – moving stairs or travelator have revolutionized the planning of buildings, and have in many ways eclipsed the role of staircases. This chapter reviews the historical development in the mechanics of shifting people up, down or through a building. The material presented owes a great deal to the help given by Kevan Goetch and Barry Wheeler from Otis Elevator plc.[1]

Readers wanting precise information on the sizing of elevators and escalators should turn to the industry itself, which operates a design service covering traffic calculation, number and size of equipment as well as giving advice on motive power. An authoritative guide is available from the Chartered Institution of Building Service Engineers. (Guide D: Transportation systems in buildings (1993).)

12.2 Historical review

The earliest recorded lifts are those described at the Colosseum where cage hoists were installed for lifting animals and gladiators into the Roman amphitheatre. The operation was effected by a simple capstan and pulley system raising a section of floor within guide rails. Treadmills are associated with medieval construction, where many examples are still working – the most famous is the trolley elevator astride a buttress wall built to raise provisions to the Abbey of Mont Saint Michel. A more conventional treadmill (*circa* 1332) exists within the loft space of Beauvais Cathedral; it hoists slates and timber from the nave floor to roof level and is in use today. (Figure 12.1*a*).

Counterbalance weights were employed with specialized hoists for lifting stage scenery in the seventeenth and eighteenth centuries. An adaption of these principles gave rise to 'chaises volante' and 'table volante'. The former was installed at Versailles by Louis XV for the benefit of his mistress, the Duchess de Chateauroux in 1744, it was dismantled in revenge by her successor, the Marquise de Pompadour. The 'table volante' enabled a complete section of a dining room floor to be raised or lowered at will to lay or clear the table and place settings. Scissor hoists were later employed for this purpose as constructed in Linderhof by Georg Dollman for Ludwig II of Bavaria (Figure 12.1*b*).

The Industrial Revolution introduced water and steam to power hoists; the

Figure 12.1 Early days of elevators
a Medieval cathedral hoists (from Gavois, J., Going Up, *Otis Elevator Company, 1983)*

most primitive form reserved for mine shafts was termed a 'man engine'. This comprised a single pole of connected timbers with platforms at regular intervals (Figure 12.1c). The up and down action was operated by a steam piston engine with the miners having to adroitly jump on and off the fixed platforms at the side of the shaft. Conventional elevators using passenger cages and guide cables were water powered in those days, the customary term for the lift had an antique quality – the 'ascending room' (Figure 12.1d). Such patterns were in common use until the 1870s and relied upon hydraulic rams to raise and lower the platform, not unlike 'Oildraulic' powered machinery today. Water powered elevators were widely employed in the apartment blocks of Victoria Street, London (refer back to plans and details in Chapter 3 Section 3.3.1). They were still found working in buildings close by Westminster Cathedral in the early 1960s, although safety regulations by that time forbade operation in frosty weather.

The revolutionary development of rack and pinion safety locks coupled with steam powered machinery entirely changed lift engineering with the following dates as marker points in the new technology.

Stairs, Steps and Ramps 261

Figure 12.1b 'Table Volante' below the dining room of Ludwig II at Linderhof, 1886. (George Dollman)

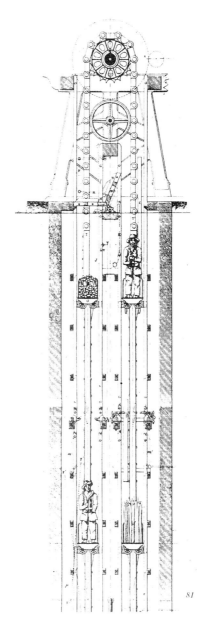

Figure 12.1c 'Man Engine' as used in Cornish tin mines. (From Gavois, J., Going Up, Otis Elevator Company, 1983)

1853 Elisher Graves Otis demonstrated the safety lift by cutting the suspension cable at the New York Crystal Palace Exhibition while a passenger on his own machine.

1857 First Otis passenger lifts fitted in a Broadway Store, New York.

1859 First Otis steam powered lift.

1865 First Edoux hydrauliic lift for the building trade.

1880 First Siemens electric lift.

1889 Otis electric lifts for the ascent to the second platform at the Eiffel Tower.

1890 Paris International Exhibition. Moving pavements and escalators demonstrated for the first time. Escalators were patented by Otis and the word registered.

Technical progress was not achieved without the dilemma of national pride. The most publicized problems were those encountered with the lifts for the Eiffel Tower.[2] The 'hockey stick' profile of the corner towers presented considerable difficulties, not least the restrictions

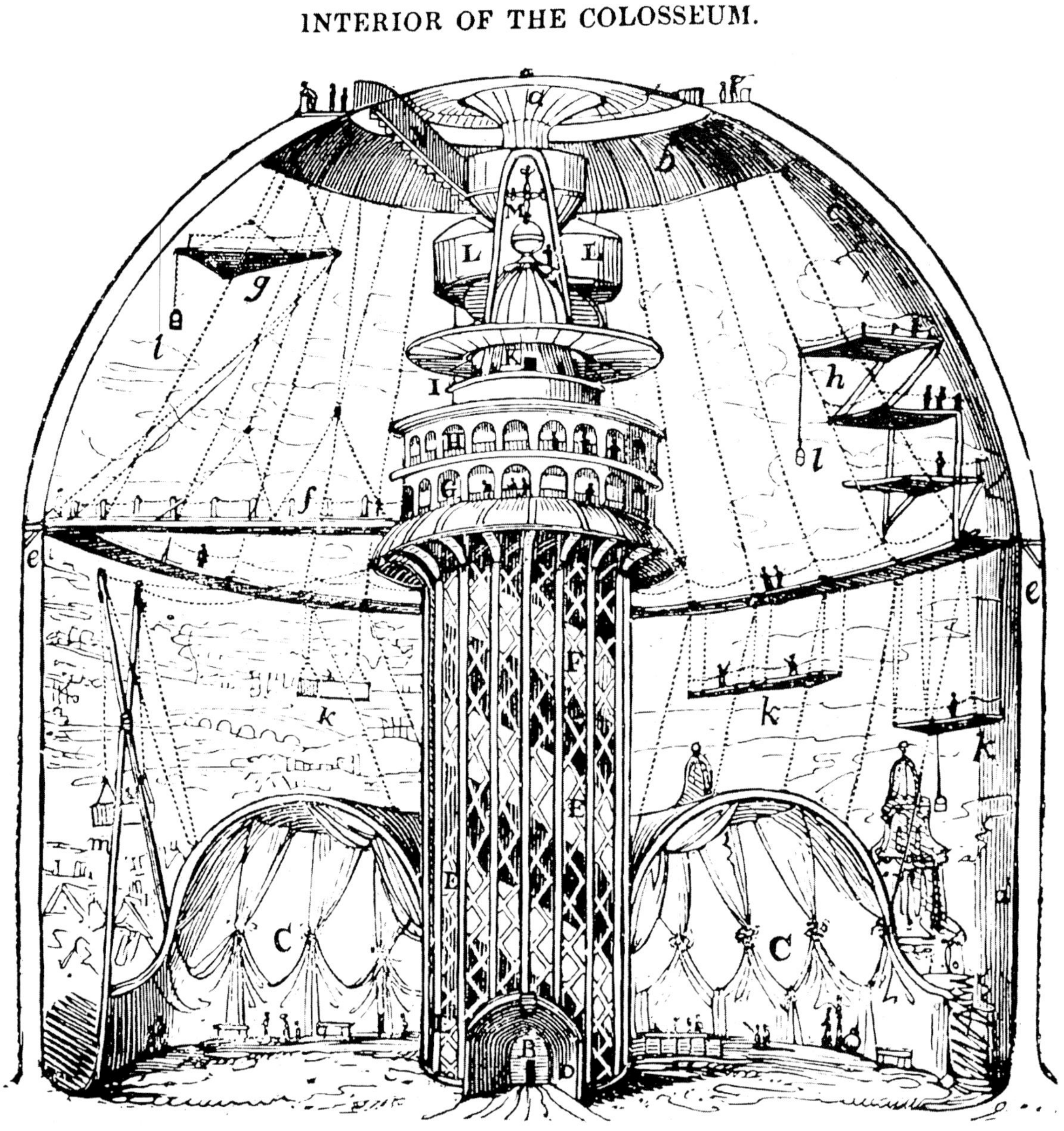

INTERIOR OF THE COLOSSEUM.

Figure 12.1d 'Ascending Room', detail of interior of the Colosseum, London, 1829. (Decimus Burton) (From Gavois, J. Going Up, *Otis Elevator Company, 1983)*

imposed by the designer who required uninterrupted views through the central space up to the second platform; visible lift overruns, machinery or cabling were forbidden between the splayed struts. Foreign firms were not permitted and French firms refused to tender for the 'hockey stick' run. The impasse was solved with three systems of ascent (one American and two French). First, a Roux-Combaluzier mechanism that depended upon articulated rods (like drain rods) that were forced up or down tubes by a horizontal steam piston engine. The noise

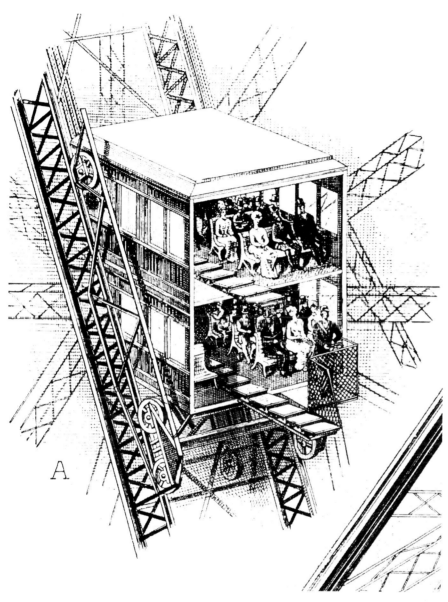

Figure 12.1e Lift shafts at the Eiffel Tower. (From Gavois, J. Going Up, *Otis Elevator Company, 1983)*

and vibration was considerable. They were replaced within a few years. The second and more difficult installation was the equipment that could negotiate both the straight and curved profile of the structure to take passengers directly to the second platform (at 116 m level). Otis devised a two decker cabin running on rail tracks with electric machinery, hydraulic gear, cables and counter balance trolleys. (Figure 12.1*e*). The ultimate 160 m to the top were operated by counterbalanced lifts, involving a passenger change at the half-way point. These were also worked by hydraulic power but to the designs of Edoux, a fellow student with Eiffel at the Ecole Centrale. There were however considerable delays,

Figure 12.1f Lift cage in the Bradbury Building, Los Angeles, 1893. (George M. Wyman) (See also Figure 4.31)

Gustave Eiffel was a fit 57 year-old when he made weekly inspections up and down the tower *on foot* in 1889. One report says that on the day of the Presidential visit the Roux-Combaluzier and Otis equipment was still not working. President Carnot, a plumper 57 years old was obliged to walk up and down to the second platform (116 m level) so as not to lose face with the engineer–owner of the Eiffel Tower and the assembled workforce. Engineer and President both wore top hats and morning coats for the inspection.

Lifts in France until this time had been developed for a rise of no more than 50 m, the transformation achieved with the Eiffel Tower (despite three different principles being employed) irrevocably altered the concept of moving people by mechanical means. The experience gained by Otis with North American skyscrapers meant that machinery could handle lift cages for 40 people at speeds of 121 m per minute at the Eiffel Tower in 1889.

The movement pattern achieved by lifts totally changed the way tall buildings could be contrived. Dreams like the 'mile high' skyscraper of Frank Lloyd Wright have still to be built but the high rise architecture that distinguishes the twentieth century from previous history would never have occurred without the pioneering work of lift engineers in the nineteenth century (Figure 12.1*f*) The escalator and travelator first demonstrated at the Paris Exhibition in 1900 have in turn transformed the handling of crowds in vast multi-storey spaces.

12.3 Lifts today

The basic types of lift can be defined by operating mode, each having their ideal application. Governing factors are the required performance in speed and capacity as well as running costs.

12.3.1 Roped systems

The majority of installations depend upon roped operations with ratios of 1:1 or 2:1 according to pulley arrangements. The most economic working is obtained with the electric traction unit mounted above the lift shaft. Mounting the traction machinery at or below the lowest floor served will reduce the shaft overrun to about 1 500 m and saves the cost of a construction at roof level. However, there are much higher running costs and greater loading on the shaft due to head pulleys. The selection of basement machinery rooms usually arises due to difficulties of accommodating structures above top floor level in conversion work or where architectural considerations limit the skyline. However, the greatest range in capacity travel and speed can be obtained with roped systems employing overhead traction units.

12.3.2 Hydraulic systems

These systems are derived from the nineteenth century but operate today with electric motors and pressurized oil. The basic provision includes a telescopic hydraulic cylinder which is housed centrally below the car or on either side. The simplest arrangement is the central cylinder with a shaft drilled down into the subsoil. Alternatives involve single or double cylinders to one or both sides of the shaft. The advantages rest with the minimum loading on the building and with the fact the machinery room can be at the lowermost level and up to 15 m away from the shaft. The limitations are speed and duty loads, say 0.63 m per second for loads around 1 000 kg. They are ideal for low rise (up to six storey) buildings with non-intensive use.

12.3.3 Scissors lifts

Mainly employed for goods handling between two floors and a descendant of those employed for the dining table of King Ludwig (Figure 12.1*b*).

12.3.4 Paternoster lifts

A design depending upon an endless chain with cabins ascending and descending continuously between winding gear. The origins are industrial, with circulating pallets within a multi-storey plant, and are also used for refuse bin circulation in flats. Paternoster lifts were popular in the immediate post-war years for serving high rise offices and public buildings with six or more pairs of cabins working side by side. Lack of safety stops in case of blockage by goods or by passengers falling across the open fronts has lead to the replacement of 'paternoster' equipment. The designs did not inspire confidence quite apart from the risk of severing leg or neck. The passenger cabins rattled and were often finished in identical, but scuffed, material to ceiling, floor *and* wall giving the impression to the uninitiated that the conveyance went round the winding wheel leaving any left standing upside down for the descent!

12.3.5 Design selection and core planning

Selection of a system and operation depends upon design studies that take into account building use and occupancy. A single lift will generally only be suitable for a modest commercial building of five storeys and then provide poor service.

Lifts are commonly grouped in pairs, fours, etc., and are better distributed either side of a lobby which does not double as a through passage, lobbies should be twice the depth of the car (Figure 12.2*a*). Modern control methods can maximize efficiency with collective calls either downwards or upwards or in both directions for the single lift or with groups and can be programmed to resolve peak traffic movements, mornings and evenings. The twin deck car is a device for speeding traffic in hotels with alternative stops. It

266 Stairs, Steps and Ramps

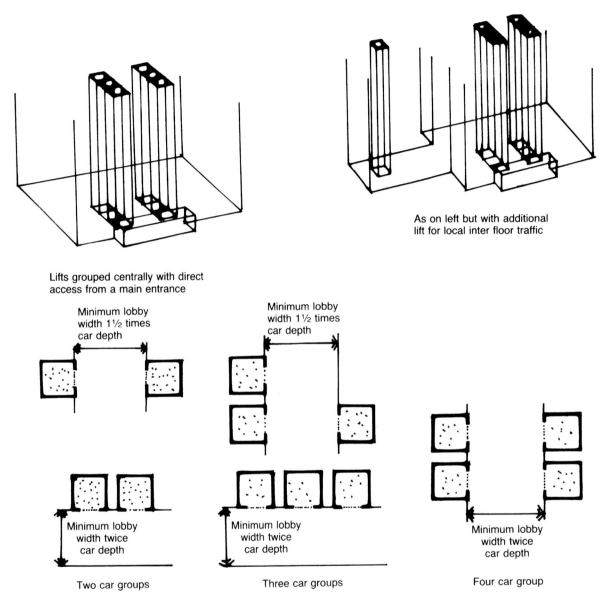

Figure 12.2 Core planning
a Lift lobbies

is also useful in tall office blocks where entries occur at both street and podium level (Figure 12.2b).

Car and landing door configuration affect efficiency, the two panel, central opening pattern being the favoured solution. Four panel designs are needed for wider, larger cars. Asymmetric plans are more space effective but slower in operation. This concluding illustration (Figure 12.2c) demonstrates the restrictions placed upon core planning with a high rise commercial building. The typical floor print in the AT&T headquarters, New York reveals the 'race track' layout around the central core, with the lift lobby serving two rows of cars and the separation of service lifts. The space of the lobby is dictated by fire regulations and also by the space needed for the machinery room at roof level. If residential floors exist these have bypass shafts and separate entries at street

Stairs, Steps and Ramps 267

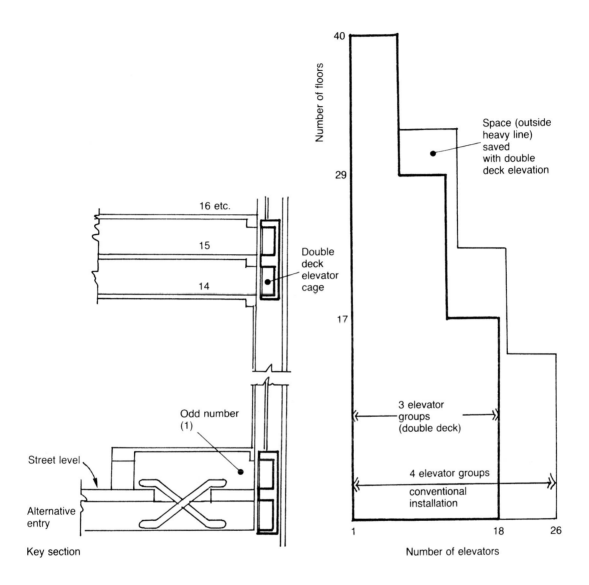

Figure 12.2b Twin deck lift cars

level. At street level the office lobby would be arranged solely for the cars and not form part of a general passageway. In developments of 50 or more storeys it is common for lift operations to be staggered, for example, the lower group to serve up to level 25, the upper group to run level 1 to 25 non-stop and then stopping for the remainder, whilst a pair of cars might run non-stop to the penthouse suite.

12.4 Escalators

One of the earliest large scale applications of escalators were those installed by Otis for the Bakerloo and Piccadilly lines of the London Underground in the early 1900s, the familiar wood panelled designs with slatted wooden treads have only recently been removed following a terrible fire at Kings Cross. The original 'moving stairs'

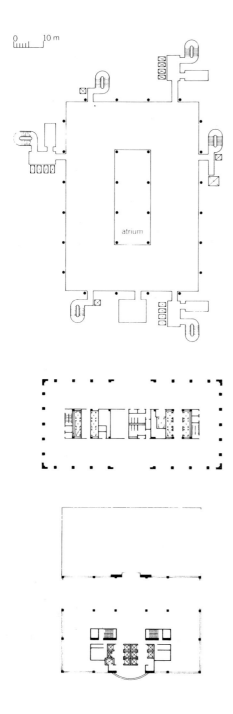

Figure 12.2c Lift core planning, flexibility in planning office space

were the pride and joy of Frank Pick, the innovative manager of London Transport from 1907 onwards,[3] the high standards of inspection and upkeep in times past meaning that fires did not occur, though ply and timber was retained for steps and their encasement. The pattern was copied in pre-war Moscow but the speeds increased from 0.75 to 1.00 m per second.

The early uses were in department stores, although the public were wary of the new invention even when geared down to 0.5 m per second. Otis were obliged to engage a wooden-legged Naval Officer in uniform to go up and down the newly installed escalators of Gimbel's store in Philadelphia, *circa* 1901. Those same 'moving stairs' had been demonstrated in the Paris Exhibition the previous year. The wooden-legged officer did further service as escalator attendant when installations were made in Paris and New York and finally in London, at Harrods store. Alternative moving stair designs were developed as 'corkscrew' forms, that were apparently installed at Holloway Road Station on the Piccadilly Line but removed due to continuous break down. Another early idea were moving stairs cum travelators for sideways access and egress that proved dangerous in use.

The 1930s Harrods escalators elevated the new concept of moving stairs to an artistic form. Working models (Figure 12.3a) were made to show the Board the dramatic idea of the new escalator hall serving all floors of the department store. The finishes in polished brass provided a light reflective surface for both day and evening enjoyment, the 'portal wall' which led to the shopping floors had varied decor identifying the levels (Figure 12.3b). Regrettably the whole interior has been ripped out. The ebullience of the designs by Louis D. Blanc[4]

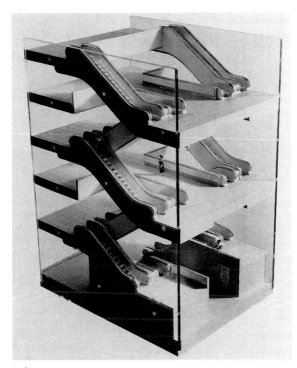

*Figure 12.3 Escalator design
a Concept model for Harrod's escalators (late 1930s). (Courtesy of Mr M. Al Fayed, Chairman of Harrods Ltd)*

Figure 12.3b Interior of escalator hall, Harrods, finished in polished brass, 1939. (Louis D. Blanc) (Courtesy of Mr M. Al Fayed, Chairman of Harrods Ltd)

can still be seen in D.H. Evans, Oxford Street, dating from the same period.

Reversible escalators were made for the underground and have a useful role in conference facilities or theatres where reverse flow is needed for intervals or at the end of a performance, significant examples are the Kremlin Palace of Congresses, Moscow (1961) designed by Posokhin and the London Theatre (mid-1960s) by Sean Kenny.

The acceptance of escalators as a normal conveyance has made immeasurable changes to the way buildings entries are laid out. The nineteenth-century idea of the spacious ground floor radiating towards grand stairs leading aloft is replaced by a multi-level vestibule. In these spaces the prime area may be below or at first floor level enabling secondary and service functions to be tucked away. The elevation to first or second floor is a common solution in commercial designs, for example hotels and offices, where ground floor space may have premium value for retailing as well as providing space for vehicle ramps and for 'porte-cochere'. These ideas can be traced to pre-war designs, a truly seminal example is the first floor banking hall reached by escalators in the P.S.F.S (Philadelphia Savings Fund Society) Building by George Howe and William Lescaze of 1932. In London, the Canadian architect, Howard B. Crane built a grander version at Berkeley House in 1937–8, now modernized but not improved.

A pair of current Chicago solutions celebrate escalators to the virtual exclusion of lifts and stairs as the first impression upon entry. A splendid example is Water Tower Place (refer back to

Figure 12.3c First stage entry to mall at Water Tower Place, Chicago, 1976. (Loebl, Schlossman, Bennett and Partners) (See also Figures 4.27a, b)

Figures 4.27*a* and *b* for upper floors). This hotel and shopping mall atrium is entered at first floor level via a special sequence, filled with a cascade of planters, water and metallic escalators (Figure 12.3*c*). The office towers of 311 South Waker Drive have a complex podium at second floor level approached by escalator halls of Cecil B. De Mille proportions. These in turn embrace a horseshoe run of food outlets and shops, in addition to a sunken conservatory (Figure 12.3*d*). All these elements effectively suppress the working core of lifts and service stairs which serve 50 or more storeys of offices above the podium to a lobby space of no significance. By contrast the latest Chicago addition to the architectural revival of 'beaux-arts' styling is a massive library that appears to be a railway station. It is in fact a new complex dedicated to reading rooms and book stacks in five storeys. The inevitable Palladian vestibule lands the visitor in a large inner hall. There is no obvious ascent in sight, no flashing lift indicators within view nor comely staircase. There is however an elegant hand painted sign

Figure 12.3d Main entry to 311 South Waker Drive, Chicago, with escalators and conservatory hall, 1990. (Kohn Pederson Fox with Harwood K. Smith)

(Figure 12.3*f*) which points back almost to the way one came in to a set of escalators, tucked away in a recess and ascending to the other floors. It does seem inexplicable that a cultural revival in the 'beaux-art' can occur without classical planning principles. The distinguished work of Garnier or Semper (reverting to Figures 2.26 and 5.3) produced entries and stairs that unfolded in a logical manner. Designer's stairs or escalators that need a sign saying in effect 'this way' are a severe disappointment.

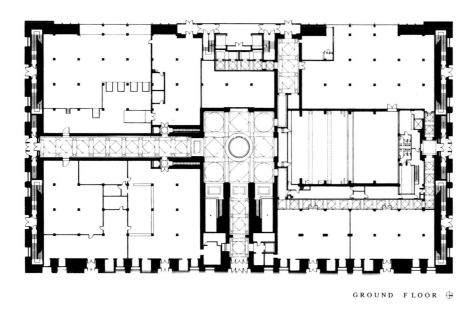

Figure 12.3e Harold Washington Public Library, Chicago, 1991. (Hammond, Beeby and Babka) Plan

12.5 Construction

Escalator construction is essentially a trussed beam to support the loop of stair-chain and treads with winding gear top and bottom. Recent changes have brought about transparent balustrades, while some designs such as the installation at Lloyds Headquarters, London have the working mechanism exposed to view within a glass sided truss (Figure 12.5p). Travelators are similar but lengths over 10 m will need supports at floor edges. There are limitations too on the rise of 6 m with inclines of 10 degrees and 12 degrees at speeds of 0.5 and 0.65 m per second respectively. Horizontal or near horizontal runs can be manufactured up to 120 m with inclines between 0 degrees and 6 degrees at speeds of 0.65 to 0.75 m per second. Foot space is needed for both types of equipment to house machinery around 2.5 m horizontally beyond the sloping part of the truss frame.

Figure 12.3f Sign to point out invisible escalators

12.6 Case studies

This chapter concludes with a review of related designs that exhibit new trends in the roles of elevators, escalators and travelators. The material is subdivided into design themes and studies in the following groups.
- A. The metal cage, glass cabin and wall climber.
- B. The glass lift enclosure.
- C. The multiple core lift and/or elevator.
- D. Travelator-scape.

Case study A: The metal cage, glass cabin and wall climber

The key reference point is the atrium in the Bradbury Building, referring back to Figure 12.1f, where the bronze caged lift car rise in the open to the office balconies. Frank Lloyd Wright makes an attractive allusion to a 'bird cage' theme in the lifts designed for the Johnson Building, Racine (Figure 12.4a). Here the circular cages run within an open framework through two storeys and are placed to give overall views of the celebrated work space with the mushroom columns. It is part of the sense of openness that the directors' rooms have clear glass between their working areas and the general office.

Case study B: The glass lift enclosure

The Asplund designed glass enclosure for the lift at the Gothenburg Court House takes minimalist ideas to their furthest

Figure 12.4a Case study A:
The 'bird cages' designed for screens and cars at the Johnson Building, Racine, 1936. (Frank Lloyd Wright)

extreme with a transparent cabin ascending within the open space to the balconies on the first and second floor. The siting is placed at the critical junction between the main entrance and the base to the formal stairs rising to the first floor courts. Figure 12.4b reveals the way the balance is struck between the formality of the judges stairs and the modernity of the lift. This design is the source of the glass jewel or watch case inspiration seen in subsequent designs by Jacobsen, Foster, Hopkins and Rogers to name but a few.

The other case study is concerned with the detailing of the glass cabin and screening of the pair of hydraulic lifts which

Stairs, Steps and Ramps 273

Figure 12.4b Case study B: Asplund's glass lifts at the Courthouse extension Gothenburg, 1937. (See also Figures 5.18b and c)

connect the basement, ground and mezzanine floors in the Sainsbury Arts Centre, Norwich completed in 1977 by Foster Associates. The first sketches by Norman Foster indicated glass walls wrapped round a minimal steel frame, with a grilled ceiling to cope with maintenance access and the housing for the door mechanisms. The final details for section and plan are given in Figures 12.4c and d. The toughened glass enclosure to both the shaft and lift cars minimize their visual intrusion into the gallery. Using hydraulic rams means that the drive machinery is concealed at basement level. The precision jigging and high quality welding in the lift cars was a result of involving a car maker with the fabrication. Steel and aluminium for the moving parts are enamelled white while the static elements are aluminium or plated steel to match the metal finishes at mezzanine level. Figures 12.4c and d

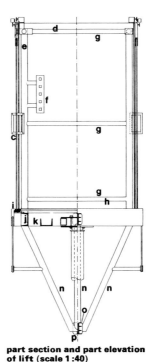

part section and part elevation of lift (scale 1:40)

Figure 12.4c Case study B: Section detail of glass and steel framing to lifts at the Sainsbury Arts Centre, Norwich, 1977. (Foster Associates)

plan section through lift (scale 1:40)

Figure 12.4d Section and plan detail, Sainsbury Arts Centre

detail the lift car as manufactured and the view in Figure 12.4e places the design within the context of the gallery space.

A similar theme of transparency is given to the new lift within the existing dinosaur display at the National History Gallery where the shaft and walkway are part of the circulation route inserted by Patrick Herron (Figure 12.4f). The theme of transparency with all the working elements made visible and the car transformed to a glass bauble is the current vogue for wall climber lifts. The precedents are worth studying. Arne Jacobsen created an open framework at the city hall of Åarhus in 1942 (refer back to Figures 5.5c). This is captured and enlarged in the designs for lift cores combined with stairs in the Institute du Monde Arabe, Paris (Figures 12.4g, h). Here the notion of disappearing reflections and transparent screening is taken to the ultimate degree.

Compliance with safety requirements can often change a remarkable experience to one that is mediocre. The polished

Figure 12.4e View of glass and steel framing within the context of gallery space Sainsbury Arts Centre

Figure 12.4f Insertion of glass lift and metal bridge within the dinosaur display at the National History Museum, 1991. (Herron Associates)

Figure 12.4g and h The notion of reflections and transparency taken to the ultimate degree. Lifts and stairs in L'Institut du Monde Arabe, Paris, 1988. (Jean Nouvel)

cages that graced Selfridges (1916) in times past now reside in the London Museum (Figure 12.4i) while the replacements are not worth recording. The same disappointment can be expressed for the Italian lifts that replaced the elegant fluted art glass work from the days of Edward Maufe's designs for Heals in the 1930s.[5]

The notable exceptions from our own time are the lifts placed in the open space encompassed by 'Le Grand Arche' Paris. The change in siting from internal to external has made a masterpiece of the composition (Figures 12.4j, k). The asymmetry and grand modelling provide a

Figure 12.4i Beaux-Arts elegance: the bronze cages and decoration at Selfridges store (1916) now in the London Museum. (Daniel Burnham & others)

276 Stairs, Steps and Ramps

Figure 12.4j The lifts that grace Le Grand Arche, Paris, 1990. (Johann Otto von Sprekelsen)

Figure 12.4k Detail view of lift construction, Le Grand Arche

sense of surprise and captivate the eye from a distance and fulfilling the visual qualities needed when close up to an overscaled building. The 'cats cradle' of cables contrasts with the panel grid façade, the movement through space to the exhibition hall above is an experience that rivals those obtained from the Otis route to the second platform of the Eiffel Tower.

Case study C: The multiple core of stair, lift and/or elevator

This selection has been culled from two pairs of buildings, each having a particular approach to the movement of people. The individual works have a distinguished pedigree, each representing a culminating point in design development with lifts and or escalators by Foster Associates[6] and Richard Rogers and Partners.[7]

The Willis Faber and Dumas Building has a superb elemental plan with the twin escalators serving the central bay of the layout. The mandatory cores with lifts and stairs are distributed at four points within the remaining areas (Figure 12.5*a*). The ascending escalators through the central bay open the departments to view floor by floor before reaching the naturally lit penthouse. The equality of treatment provides an egalitarian work space; there are equal compensations for all with the roof garden restaurant and leisure facilities within the compass of the escalator bays (Figure 12.5*b*).

The Hongkong and Shanghai Bank stretches the ascending spaces of Willis Faber into a vertical stack of floors that rise 40 storeys (Figures 12.5*b*, *c*, *d*, *e*, *f*). The subdivision of a banking headquarters, unlike insurance, falls into three zones, the public domain of street and trading floors, (Figure 12.5*e*) the working areas (Figure 12.5*f*) and the zones where privacy and security are most important. The public domain is served by a pair of escalators from the plaza to the main public floor, known as Level 3; the angle between the flights and the slope were determined after discussions with the Fung Shui adviser. A further pair climb to the secondary banking level (Level 5). Inter-floor escalators cope with office traffic from Level 10 through to Level 35. (Figure 12.5*d*). The structure of the skyscraper amounts to four pairs of trussed columns that have adjoining service cores of fire stairs, lifts and lava-

278 Stairs, Steps and Ramps

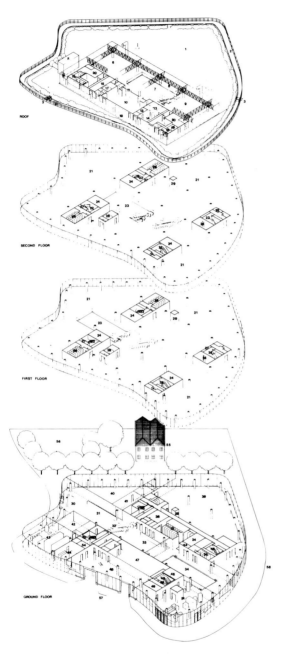

Figure 12.5 Case study C:
a Willis Faber and Dumas Office Building,
Ipswich, 1975. (Foster Associates) Isometrics

tories (Figure 12.5c). The lifts have access to all floors and are distributed to both sides of the floor spaces served. Four double height zones exist within the tower connected by high speed lifts to

Figure 12.5b Interior view of escalators within central bay of plan, Willis Faber and Dumas office building

form secondary entrances. These also provide the refuge areas required by the fire regulations in Hong Kong, a further advantage being the division of the façade. The most memorable impression of the composition is the entry from the plaza level with the gentle angle of the escalators penetrating the glass membrane floor of the atrium space, seen at its best through the length of the plan. The integration of movement spaces within the structure is helped by the logical disposal of stairs and lifts to the edges of the spaces served.

The Centre Pompidou in Paris arose from the International competition in 1971, when the design by Piano and Rogers was selected as the prize winner. It is said that the concept of the major

Stairs, Steps and Ramps 279

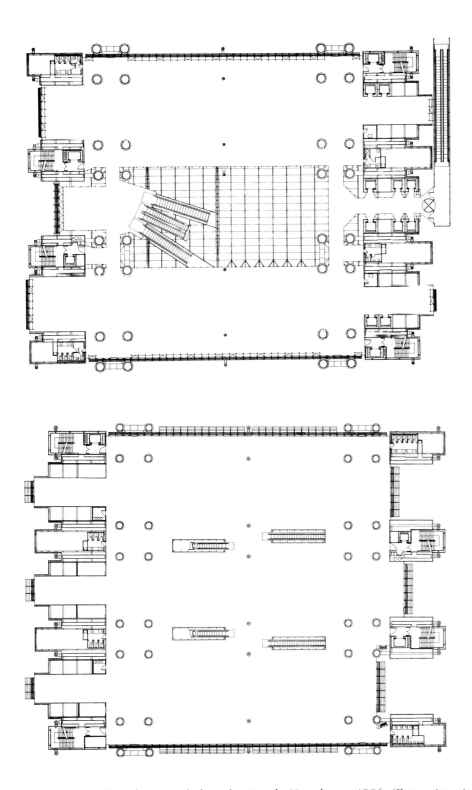

Figure 12.5c Hongkong and Shanghai Bank, Hongkong, 1986. (Foster Associates) Plans

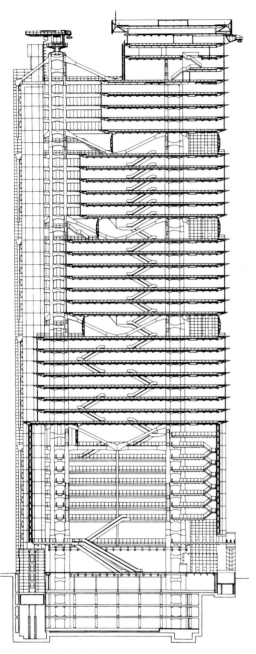

Figure 12.5d Movement diagram, Hongkong and Shanghai Bank

space with servant spaces to either side for entrances, stairs and services was the idea of Peter Rice. The public face of the Centre Pompidou is on the west side towards Place Beaubourg, which has the main entrance from the sunken level of the square. The entry to the Centre follows an anticlockwise pattern with internal escalators that take visitors to the upper mezzanine level where there are galleries and circulation space, these in turn lead round to the most popular feature, the external escalator promenade to the principal gallery and museum space on the third, fourth and fifth floors (Figure 12.5*b*). For many visitors the sheer delight of the building is the method of circulation through the transparent escalator 'tubes' (Figures 12.5*g*, *h*, *i*). The very popularity of the exhibitions has raised the issue of increasing the number of 'tubes'. Crowd control today means queues. The spacious entries have had to absorb the paraphernalia of barriers and security checks. Re-ordering the entry arrangements and an external canopy to protect the waiting crowds would solve the present bottle neck when popular events occur. However, once entered upon 'Le grand escalier mouvant' all is forgiven, the movement, the excitement of the structure sliding past and the vision of Paris is spectacular. It is little wonder that Centre Pompidou has more visitors than Gustaf Eiffel's famous tower.

Some years ago a pair of 1:50 models stood side by side in the lobby to Rogers office: the Centre Pompidou next to the Lloyds Building. The two designs are ten years apart but represent a great similarity in the separation of building roles. Both have servant spaces clustered to the edges of the main activity areas, those master spaces enjoy clear volumes to provide total freedom in use. The Lloyds Building has a range of sophisticated options, a trading space that can envelop the whole central volume, commercial

Figure 12.5e Escalators from street to podium, Hongkong and Shanghai Bank

Figure 12.5f Escalators for traffic within building, Hongkong and Shanghai Bank

offices that might diminish or grow again according to market forces and finally the headquarters organization in the basement and penthouse suites. The servicing of Pompidou is handled like industrial plant of an airport or railway station. It is sturdy and robust like street furniture. Lloyds could be described as a 'machine for working in', a term used by Peter Buchanan in the *Architectural Review* (October, 1986). The lifts and stair pods, although external in form, are part of the interior experience (Figures 12.5*k, l*) and finished to Saville Row standard. The detailing of the lifts with stainless steel and glass are for 'external' wall climbers with the working components increased in specification to match the exterior climate. The critical zone between car and landing gate received special attention. A new form of retractable weather proof seal was developed by the Express Lift Company. The floors of the cars were also placed in isolation to the chassis to overcome wind turbulence. The actual enclosure was a toughened glass box made with silicone bonding. External

282 Stairs, Steps and Ramps

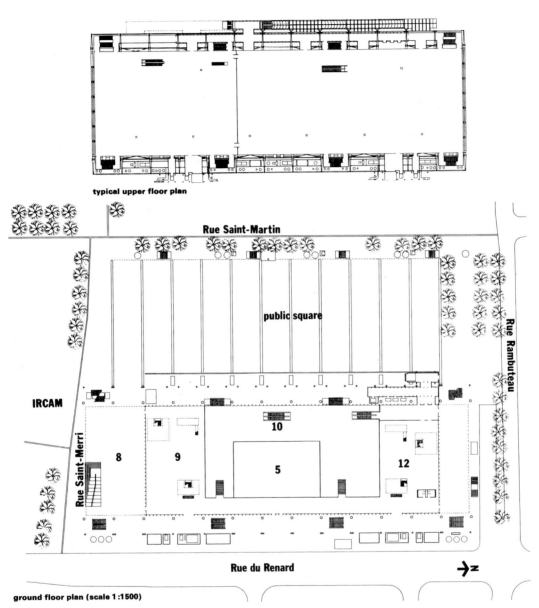

Figure 12.5g Centre Pompidou, Paris, 1977. (Renzo Piano, Richard Rogers) Plans at street level and sixth floor gallery level.

wall climbers received considerable criticism at the design stage but they have operated without breakdown. The concept, though extravagant, provides exciting views that equal those from the prominent 'escalier mouvant' at Pompidou. The stair pods are clad in stainless steel to conform to the dog-leg profile, internally the landings and stair waists are concrete with raised landings and tread/riser components in extruded aluminium (refer to Figure 12.5n for

Figure 12.5b View of escalators externally, Centre Pompidou. (Courtesy of Martin Charles)

284 Stairs, Steps and Ramps

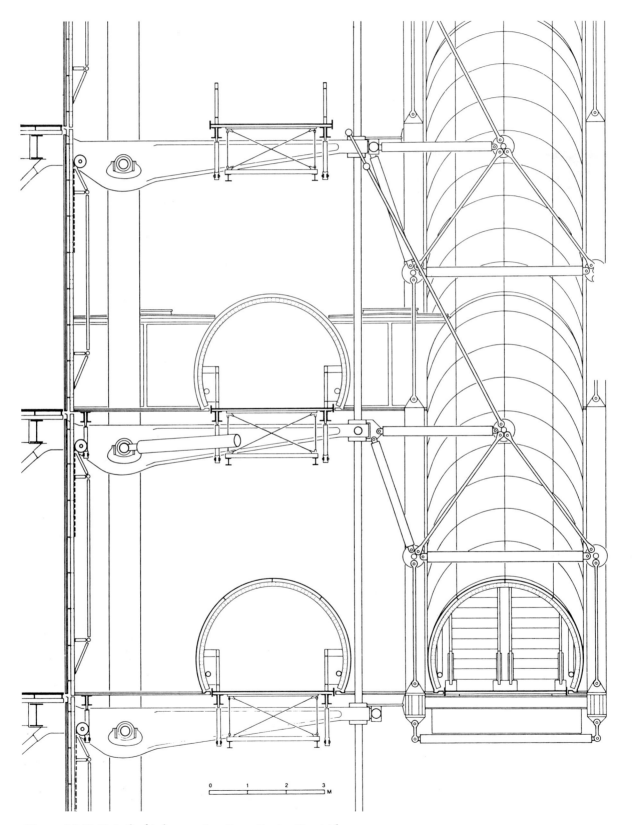

Figure 12.5i Detail of tube construction, Centre Pompidou

Stairs, Steps and Ramps 285

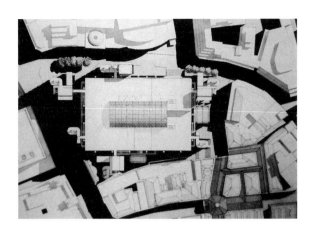

Figure 12.5j Lloyds Building, London, 1986. *(Richard Rogers and Partners)* model

Figure 12.5l View of wall climber lifts

constructional detail). The real triumph of Lloyds is the flexible interior volume bridged by cross-over escalators that enable the users to trade throughout the core of Lloyd's vertical room. The spacial arrangement is not obvious at first impact since the escalator bridge runs the shortest

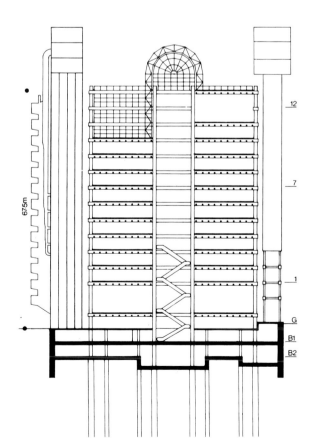

Figure 12.5k Section

Figure 12.5m External view of staircase towers

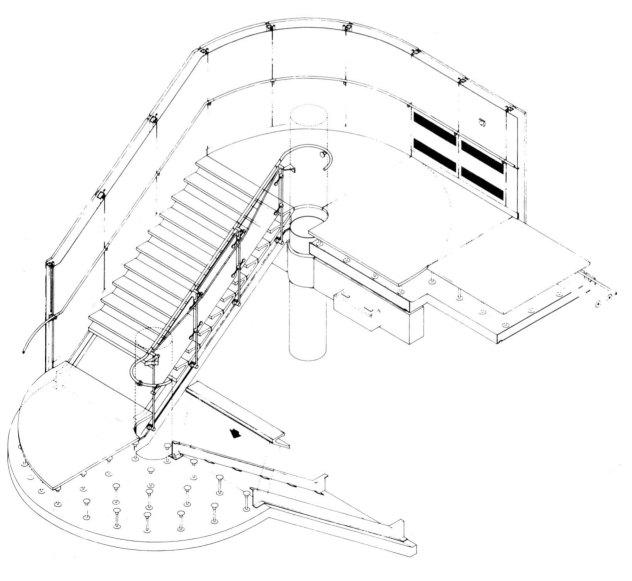

Figure 12.5n Detail construction of stairs

Stairs, Steps and Ramps 287

Figure 12.5o *General view of escalators within the multi-storey 'room', Lloyds Building*

direction unlike the public entry to the Hongkong and Shanghai Bank. In use however, the cross-over effect has the same beneficial use as the ascending form of Willis Faber. The working space can be surveyed and the subdivision made on egalitarian terms. It is as if the Frank Lloyd Wright's inspirational Larkin Building had been brought up to date by mechanical means.

Case studies D: Travelatorscape

The major application of travelators is in airports but few of these experiences are memorable. Gatwick certainly provides a convenient ride through the neutral scenery bequeathed by BAA and YRM (Figure 12.6a). The dramatic system adopted by

Figure 12.6 *Case study D: Travelatorscape a Travelator Foyer, Gatwick Airport, 1980. (YRM)*

Auketts at Manchester leads the travellers through a space-age glass and aluminium tube to connect the rail terminal to the airport (Figure 12.6*b*). Other devices in Europe include travelator tunnels that dip and climb to prevent boredom as at Charles de Gaulle, Paris. It is this airport that placed satellite escalators within a globe-like sphere that shuttles travellers off to their various departures (Figure 12.6*c*). The aesthetic delights within 'the shape of things to come' are not matched by creature comforts. It is difficult, in fact, at Charles de Gaulle airport to find a glass of water once through the space age scenery. It does seem that the future of 'Travelatorscape' should provide enhancing experiences in line with the moving girdle at O'Hara which runs half a mile with suitable breaks through a typical American Mall. A route that is lined like a modern day Aladdin's cave with drink and food outlets, fortune tellers shops, shoe shiners and every conceivable service in washrooms.

The serrated tread has enabled airport and supermarket trolleys to be conveyed safety on inclines up to ten degrees, the most stylish example are the connecting travelators within the glazed foyer

Figure 12.6c Satellite Globe, Charles de Gaulle Airport, 1970. (Courtesy of the Architectural Association)

Figure 12.6b Manchester Airport's new 240 m travelator. (Courtesy of Paul Miller)

designed by Grimshaw for Sainsbury at Camden Town (Figure 12.6*d*).

The same designer has utilized travelators to move visitors through the British pavilion at the Seville Expo. The installation permits a regular movement of people to save hold ups and to reduce queuing time. The up and down sequence occurs along the long frontages of the open plan and enables the exhibition spaces to be fully enjoyed. The internal structure of the pavilion has six bays of structural columns. These are grouped in three positions to maximize the space, the travelator trusses span between support arms jettied off the columns (Figures 12.6*e*, *f*). The constructional details are

Stairs, Steps and Ramps 289

Figure 12.6d *Travelators within glazed foyer, Sainsbury's Camden Town, 1989. (Nicholas Grimshaw and Partners)*

Figure 12.6e *Travelators within the British Pavilion, Seville, 1992. (Nicholas Grimshaw and Partners)*

printed with permission of the *Architects Journal*.

References

[1] *Going up. An informal history of the elevator from the Pyramids to the present.* Jean Gavois, Otis Elevator Company, 1983.

[2] *The Eiffel Tower*, Joseph Harriss, Elek Books Ltd., 1975.

[3] *A biography of Frank Pick*, Christian Barman, David and Charles, 1979. Pick's authoritative regime included personal inspections of stations and escalators at 2 o'clock in the morning!

[4] Louis D. Blanc (1877–1944) (no relation to the present author) carried out extensive works for Harrods 1920–30 and later for the escalator hall in 1939.

[5] Pre-war Heals is well described in the *Architectural Review*, complete with details of the 'art glass' work to the designs of Maufe.

[6] Further reading on the two Foster designs selected refer to Foster Associates *Buildings and Projects*, Volumes 2 and 3.

[7] For further reading on Rogers designs: 'Pompidou', *Architectural Review*, May 1977. 'Lloyds Building', *Architectural Review*, October 1986. See also *Richard Rogers. A biography*, Bryan Appleyard, Faber and Faber, 1986.

290 Stairs, Steps and Ramps

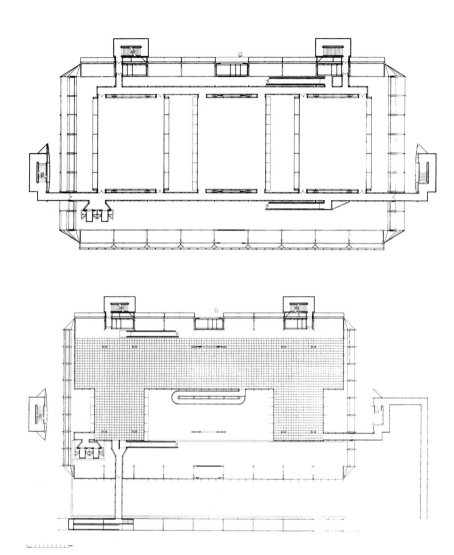

Figure 12.6f Key plans at ground and upper floor, British Pavilion

13 International case studies

The title could read 'in the last twenty years', except for the inclusion of Arne Jacobsen. The purpose of this chapter is to present outstanding designs where stairs and/or escalators and lifts play an important part; the listing is in alphabetical order of the designer. In earlier chapters, case studies and details have already drawn attention to recent work abroad, reference to the international index will expand matters.

It is not realistic to isolate two decades from the rest of the twentieth century. A group of pictures of other famous modern stairs are included as a reminder of sources, since many have some connection with the case studies which follow.

Figure 13.0b Reinforced concrete entwined ramps at the Penguin Pool, London Zoo, 1934. (Lubetkin, Drake and Tecton). The engineering work has been honed to perfection by Ove Arup with a 'V'-shaped slab which hides the true thickness of the beam element. The geometry of the two curving ramps within the armature of the enclosing walls provides a perfectly resolved geometry of ascending forms within the enclosure

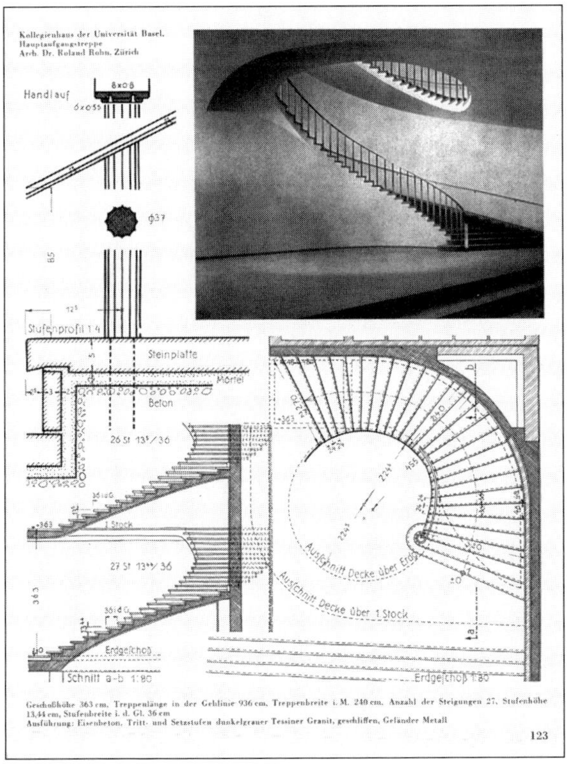

Figure 13.0a Reinforced concrete stairs, Basel University, 1937–39. (R. Rohn). The utmost refinement has been applied to the profile of the concrete work with a curved chamfer to the string to give the least dimension of the waist. The long rythmic sequence of risers is matched by robust bronze balusters

Figure 13.0c Steel and concrete principal staircase at the De La Warr Pavilion, Bexhill on Sea, 1935. (Mendelsohn and Chermayeff). The stair is placed at the turning point of the plan and celebrates the cardinal movement from floor to floor, within the building. The geometric elements include curves from the stairs and glazed windows with the armature formed by a central lighting column

Figure 13.0e Composite cable and timber steps, Research House, Silverlake, Los Angeles, 1933, rebuilt after fire in 1964. (Richard Neutra). An interesting asymmetric structure with cable work balanced by off-centre binder. The structure is further dramatized by a mirror back drop to the spine wall of the house

Figure 13.0d Suspended steel stairs, Wingspread, Racine, Wisconsin, USA, 1937. (Frank Lloyd Wright) The Wrightean concept of suspended flights had been developed for Falling Water in 1935 with steps that hovered over the cascade. The same constructional ideas are used for stairs that floated over the roof terrace at Wingspread with steel rods and pre-cast planks swung below the concrete floor slabs. The Falling Water staircase was swept away in a storm and has been replaced

Stairs, Steps and Ramps 293

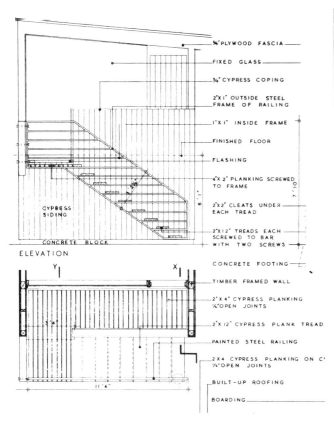 

Figure 13.0f Steel ladder support for plank treads at Exhibition House, New York, 1949. (Marcel Breuer). The ladder frame appears often in Breuer designs, the reason is a desire to combine the horizontal emphasis of the treads with the supporting structure. Close vertical bars restrict visual connections at oblique angles and blur into a maze of metalwork. Horizontal bars can however relate to adjacent landing railings and to fenestration patterns

Figure 13.0g Toughened glass and steel, RAC Building, Milan, Italy, 1956. (Gio Ponti). A complete structure in glass, laminated timber and steel which forms a viewing platform and decorative end wall to a banking hall. The landing decks contain cantilever steels from the reinforced concrete wall but the lines of handrails and strings suggest a free standing geometry in the spirit of Mondrian

Case study 13.1 University Library, Eichstatt, Bavaria (1987)

Architect: Behnisch and Partner

The pin wheel plan opens in five directions, with the primary axis from the northern loggia. The ascending stairs and lift are placed left and right of the entry with the staircase splayed towards the principal offices in the 'L'-shaped east wing. The open galleries at first and second floors give views to the library and study spaces laid out below. Those spaces in turn have circular accommodation stairs that connect with carrels or study areas on south facing balconies. Escape stairs are placed at the periphery leading to the ground or to a roof deck. The quality of stair design is not simply the choice (there are thirteen in number) but the disposition to provide control within the library and working areas. The detailing in steel tube, rod and mesh provides a light counterfoil to the generous glazing and become attractive eye catchers amongst the grid of shelving partitions and library furniture.

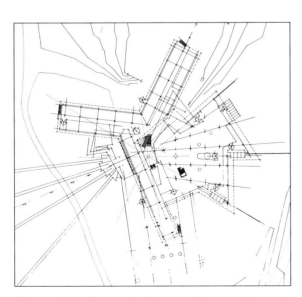

Figure 13.1 University Library, Eichstatt, Bavaria, 1987
a Key plan for ground floor

Figure 13.1b Spiral stairs used as eye catchers within the working area

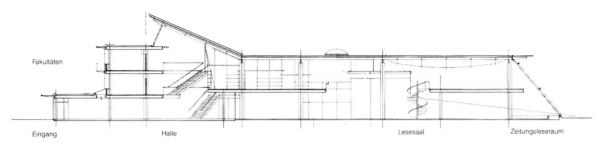

Figure 13.1c Ground floor

Figure 13.1d General view of main entrance axis with principal stairs

Figure 13.1e View from balconies towards stair and lift

Case Study 13.2
Communication Tower, Tibidabo, Catalonia (1992)

Architect: Sir Norman Foster and Partners

Structural Engineers: Ove Arup and Partners

The whole structure and the detailing has been refined to the minimum, the basis is a concrete tubular column stiffened by three steel trusses set at 120 degrees to each other. The ancillary structures are also steelwork, for instance the twelve 'technical' floors, the escape stairs and the track for the wall climber lifts. The spare elegance of the architectural engineering is superbly illustrated by the detailing of stairs and lift cage.

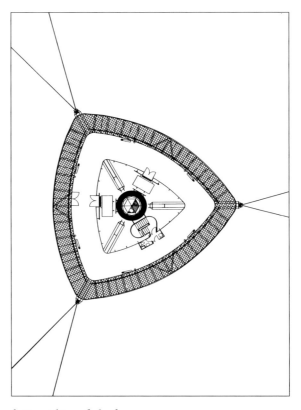

Figure 13.2 Communication Tower, Tibidabo, Catalonia
a Detail view of stairs and lift car

b Key plan of shaft

Figure 13.2c General view of tower to show staircase

Case Study 13.3 Rødovre Town Hall, Copenhagen (1956)

Architect: Arne Jacobsen

This superbly crafted piece of metalwork, has 24 mm rods to suspend the weight of the dog-leg stairs and landings. The strings are cut from 50 mm plate to a uniform 100 mm face on elevation. These follow the shape of risers and treads and mask the ends of the landing and tread trays.

The stair is placed dramatically at the rear of the main entry hall. It is seen to its best advantage outlined against the principal window that is full height in the three storey hallway. The skeletal structure, the rhythm of stepped strings combined with open treads and the elemental handrails present the most elegant and minimal silhouette.

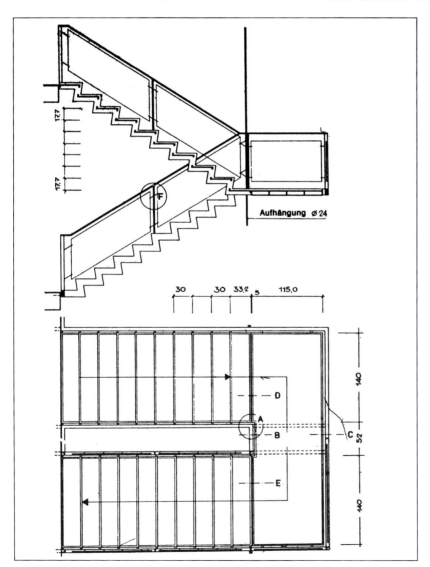

Figure 13.3 Rødovre Town Hall, Copenhagen, 1956
a Perspective. (From Pracht, K., Treppen, *Deutsche Verlags-Anstalt, 1986)*
a Constructional detail for plan and section (from Pracht, K., Treppen, *Deutsche Verlags-Anstalt, 1986)*

Stairs, Steps and Ramps 299

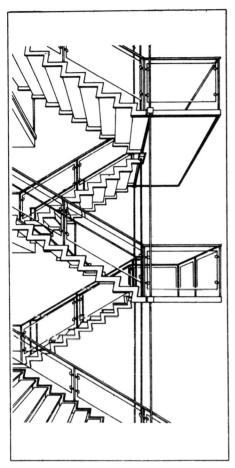

Figure 13.3b Perspective. (From Pracht, K., Treppen, Deutsche Verlags-Anstalt, 1986)

Figure 13.3c General view

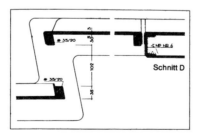
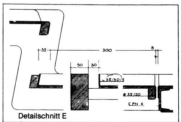
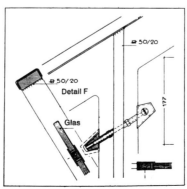
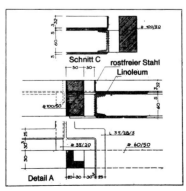

Figure 13.3d Details. (From Pracht, K., Treppen, Deutsche Verlags-Anstalt, 1986)

Case study 13.4 State of Illinois Center, Chicago (1985)

Architect: Helmut Jahn and Partners

This building is totally different from any other building in downtown Chicago. It follows the grid iron pattern on two sides in contrast to curving façades which embrace the entrance and inner circular atrium. The atrium is fully glazed externally and penetrates the total height of the offices. A cluster of wall climber lifts are the main focus within the inner space of the atrium serving the circular galleries of each floor. Accommodation and escape stairs run in a zig-zag pattern down the balcony face, contributing an expression of activity and movement within the sixteen storey volume. The lowermost floors contain a food hall and lead to the subway system. It is one of the most remarkable public buildings in the USA and particularly in the way that changing technology permits lifts and stairs to assume significance within an open interior.

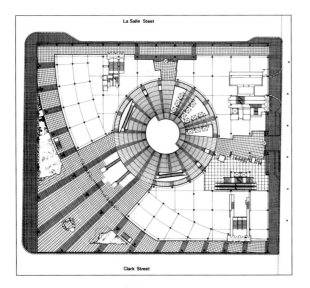

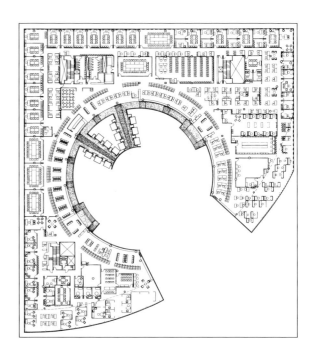

Figure 13.4 State of Illinois Center, Chicago, 1985
a Key plan

b Upper floor plan

Case study 13.5 Decorative Arts Museum, Frankfurt am Main, Germany (1984)

Architect: Richard Meier and Partners

The museum is organized astride two cross routes or internal streets to the group of buildings. The east–west path forms an open way and connects the park entrance to the undercroft of the galleries and the main doors. The north–south paths are duplicated, first the internal sequence of spaces within the museum of ramps and passages and second a footpath route that joins the riverside to the park. The internal

Figure 13.4c Detail view of stairs

Figure Figure 13.4d General view of atrium drum with wall climber lifts and stairs

Figure 13.5 Decorative Arts Museum, Frankfurt am Main
a Promenade in space

sequence is at the heart of the composition and comprises three levels with a ramp, which, as in Villa Savoye, takes over a critical role in exploring the interior. The ramp extends into a long gallery, spanning like a balcony throughout the plan at first floor. It also provides a landing space to receive the alternative vertical circulation of lifts and stairs. In Corbusian terms it is a perfect promenade in space and enables the art objects to be viewed in changing perspectives upwards or downwards and within differing settings.

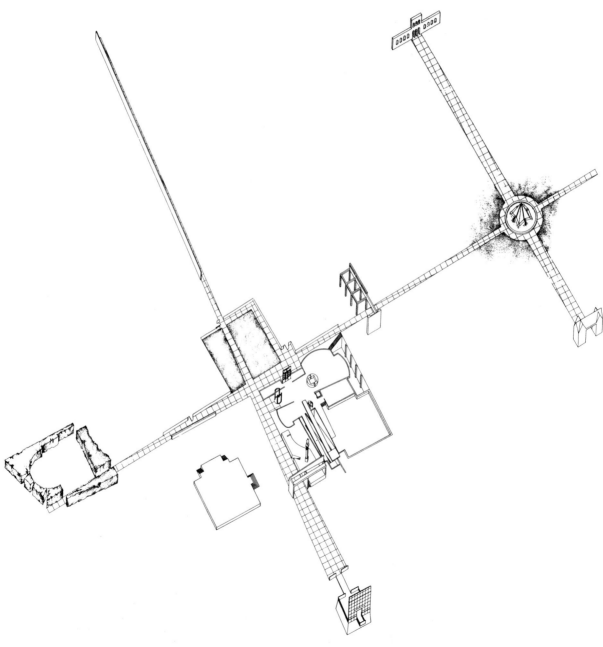

Figure 13.5 Decorative Arts Museum, Frankfurt am Main
b Relation of ground floor and site

Stairs, Steps and Ramps 303

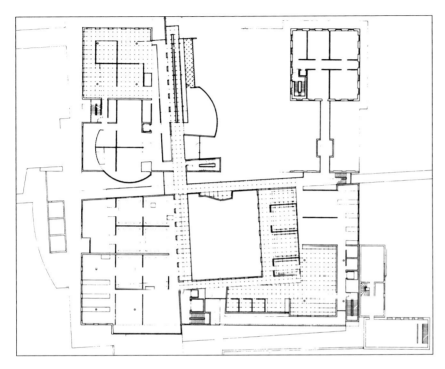

Figure 13.5c Layout plan for first floor

Figure 13.5d View of ramp from landing

Figure 13.5e View from ramp at mezzanine

304 Stairs, Steps and Ramps

Case study 13.6 Le Grand Louvre, Paris (1989)

Architects: Pei, Cobb, Freed and Partners

The Grand Projects in Paris have led to a number of outstanding designs: none more controversial or spectacular than the glazed pyramid that forms the portal to the new sunken entrance of the Louvre Museum. The popularity, as with Centre Pompidou has produced the untidy sprawl of railings to give crowd control in the old courtyard. However, once the security check is passed, the public freely descend into the new vestibule. The descent can be made by hydraulic platform or else by a gracious spiral which turns around the lift drum. The mechanics of the design are sublime with the open platform elevated on a stainless steel cylinder which descends into the ground.

Figure 13.5b View with lift drum being lowered

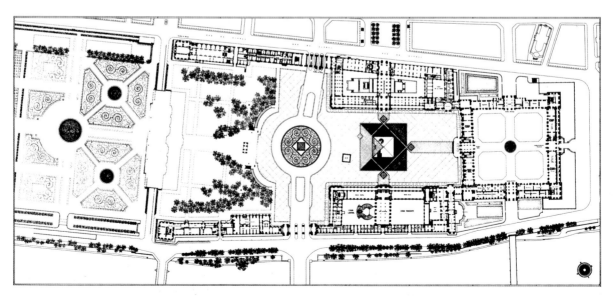

Figure 13.6 Le Grand Louvre, Paris, 1989
a Layout plan at ground level

Stairs, Steps and Ramps 305

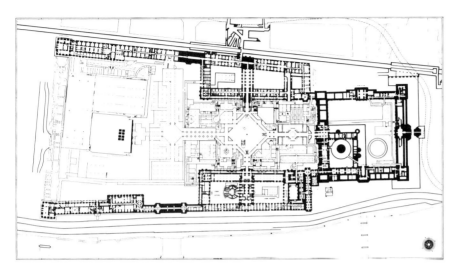

Figure 13.6c Layout at lower vestibule level

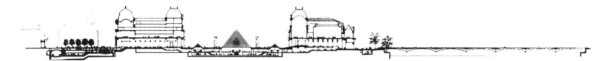

Figure 13.6d Key cross section

Figure 13.6e General view of stair and platform of hydraulic lift. (Courtesy of Koji Horiuchi)

Case study 13.7
Staatsbibliothek Preussischer Kulturbesitz, Berlin (1978)

Architects: Hans Scharoun and Edgar Wisniewski

The library composition allows for the free disposal of largely rectangular spaces set between irregular volumes. The irregularity is dictated by the varied traffic circulation with principal, secondary and incidental movement patterns. The theme is today demonstrated in the work by Behnisch. In the case of Scharoun, the expression of movement in the plan spaces is a feature since his earliest designs. At the State Library Berlin, the stairs and changes of level articulate the plan. In visual terms they are highlighted to signal the nodal zones which separate the working areas. Detail is varied and the language of scale, enclosure or openness of staircase adjusted to the role required.

Figure 13.7b General view across main library

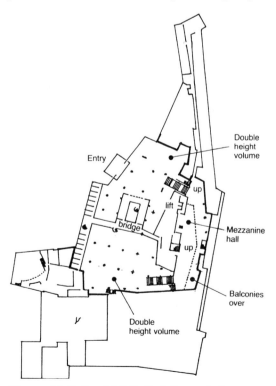

Figure 13.7 Staatsbibliothek Preussischer Kulturbesitz, Berlin, 1978
a Key plan at first floor

Figure 13.7c Detail of stairs on main approach from entry

Stairs, Steps and Ramps 307

Figure 13.7d *Individual stairs within mezzanine hall*

Case Study 13.8 Hong Kong Club (1984)

Architect: Harry Seidler and Partners

The club premises are constructed as an independent building below a 22-storey tower block. The main social area occupies four floors, including a roof terrace and garden in the space below the office tower. Twin cylinders in reinforced concrete contain enclosed lifts and open spiral stairs with access to balconies. They are planned as extended dining areas or as public areas with views down a central

well. Functional requirements concerning means of escape and the tower lifts are sited to one side and shared with the offices above. There is a freedom of form and a spatial quality relating to the soffit and ascending line of the balustrade. All this is very much in the tradition of Marcel Breuer, with whom Seidler collaborated on a number of projects.

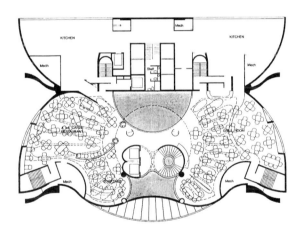

Figure 13.8 *Hong Kong Club, 1984*
a *Key plan for second floor*

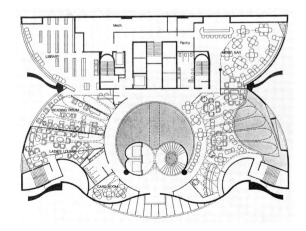

Figure 13.8b *Key plan for third floor*

308 Stairs, Steps and Ramps

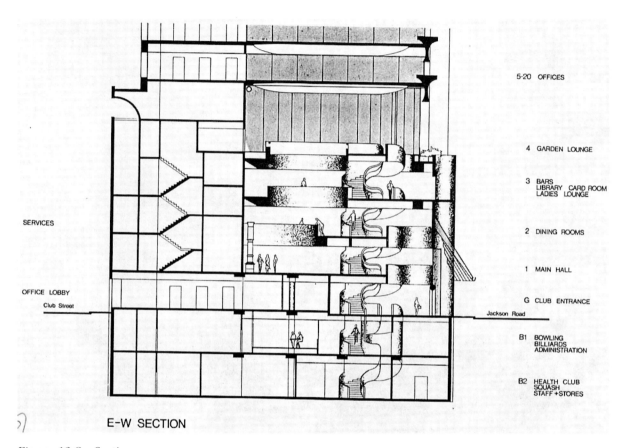

Figure 13.8c Section

Figure 13.8d General view of stair

Figure 13.8e Detail view

Case study 13.9 SAS Headquarters, Stockholm (1988)

Architect: Niels Torp

The generation of the layout can be compared to a 'High Street', with the various buildings placed freely together and the space between roofed over. The primary circulation occurs at ground level with stairs and lifts forming three meeting points. These relate to social areas which serve the whole community at the headquarters of SAS. The client stated that he wanted an architecture of interconnections, of networking and a place where people could meet spontaneously. The 'street' and its interaction with vertical circulation occurs within the social area (restaurants, seminar rooms and library) or club facilities (sports, gym, pool, clubrooms and auditorium) or at the entry point (banking and shopping). The upper floors are broken into five pavilions, each with extensive internal courts that occur along the circulation paths. The lifts and stairs rise within the open 'street' to bridges connecting the pavilions. The ideas have been tried before by Hertzberger in the Centraal Beheer, Apeldoorn and the Ministry complex in Den Haag, but Niels Torp has achieved a more eloquent expression with a finer palette than various shades of concrete.

310 Stairs, Steps and Ramps

The light airy qualities of steel and glass in balcony, lift and curving stair are in direct lineage to Asplund at the 1930 Stockholm Exhibition.

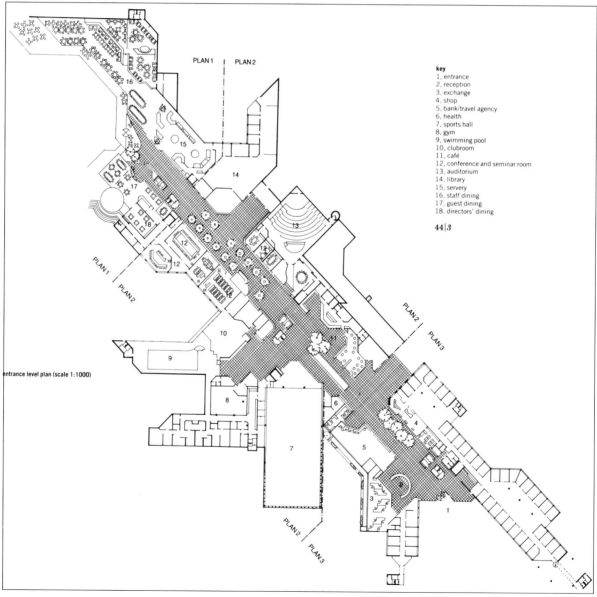

Figure 13.9 SAS Headquaters, Stockholm
a Ground floor plan

Stairs, Steps and Ramps 311

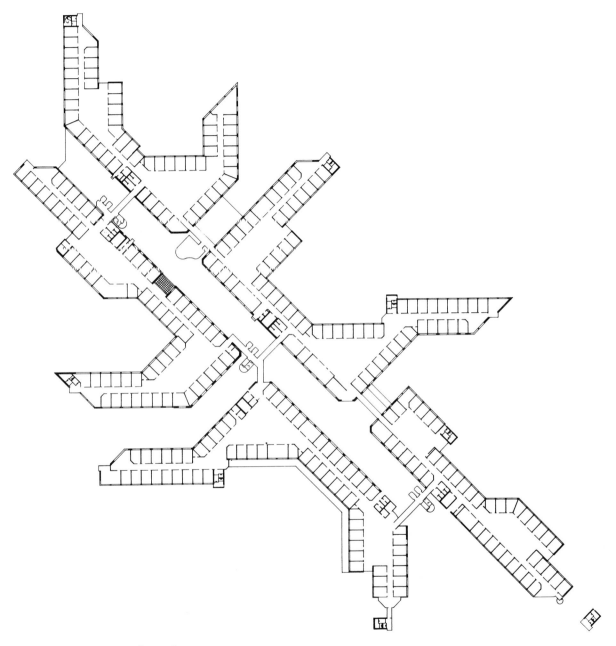

Figure 13.9b Upper floor plan

312 Stairs, Steps and Ramps

Figure 13.9c General view of street, with stairs and social areas

Figure 13.9d Detail view of balconies and lifts

Case Study 13.10 Ontario Place, Toronto (1971) and Eaton Center, Toronto (1971)

Architect: Zeidler Roberts Partnership

The touchstone in Eberhard Zeidler's work is the fine scale of detail, especially at the first point of contact with the building – the portal and its platform, or balcony, the stair, and the fit of lift and escalator. The language of detail is fully developed with Ontario Place, a marina and large urban park which fronts the lakeside at Toronto. The connecting decks have promenades and stairs of ship-shape quality.

The same finesse is seen at the Eaton Center, a rebuilt area of five city blocks astride a covered mall with three levels

Figure 13.10 Ontario Place, Toronto
a General view

Figure 13.10b Detail of ship shape stairs

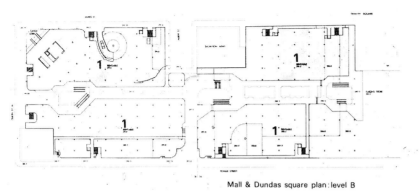

Figure 13.10 Eaton Center
c *Layout plan (street level). (From* Process Architecture, *No. 5, 1978)*

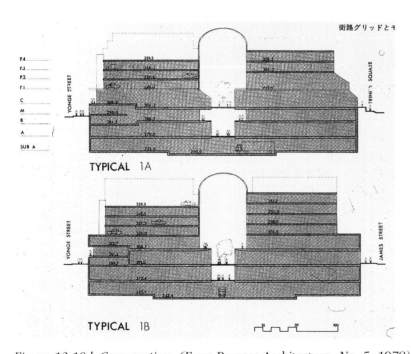

Figure 13.10d Cross section. (From Process Architecture, *No. 5, 1978)*

of pavement. The section has outward set backs to give good vision lines across the new pedestrian street. The sequence is broken at four staircase nodes, where each place is widened to form a public area with seating, plants or fountains. There are escalator connections upward to the higher level shops and downwards to the subway system. Stylish stairs also occur as part of the promenade network throughout Eaton Center. The provision of façade and promenade is in the tradition of Egon Eiermann where Zeidler was designer in the late 1940s.

316 Stairs, Steps and Ramps

Figure 13.10e General view of promenades and stairs

Figure 13.10g Detail views

Figure 13.10f View down to 'places' created at node points

14 Case studies in the UK

These case studies have been pruned to fifteen in consideration of the numerous designs already featured in the philosophic and practical chapters. The selection has also to do with availability of material, since designers are often reluctant to furnish constructional detail; the final list has a number of my favourite staircases. The subdivision accords to the method of construction with designers in alphabetical order. Special emphasis has been placed upon those designs that advance the art of construction beyond refining the traditional approach. The pictures of the steel plate treads and risers at a Smithfield restaurant are included as a demonstration of really inventive design. The inherent stiffness in short cantilevers of plate steel is used to maximum effect, the assembly depends upon masonry bearings and is without welds, a simple and effective detail. Apologies are given to those architects whose work is not included, but there had to be a balance within the selection. Metal, steel and glass stairs have attracted a lot of attention in the past decade and could form a 15-page appendix.

Timber, also metal and timber stairs

Case study 14.1 'Oakyard', Blackheath (1992); Ashland Place, Marylebone (1992)

Architects: Allies and Morrison

Engineers: Whitby and Bird (Blackheath and Ashland Place) Price and Myers (Stephen Bull)

Three examples are taken from the folio of this practice since the designers have a passion for the stair seen from all sides. The underside is as important as the top side. The first design is an elemental assembly in timber for loft stairs and shows the way the propped strings rest via metal buffers on a simple plinth. The second design occurs within the same newly built house and re-establishes the sculptural ideas developed by Adolf Loos with the sculptured steps of the Muller House, Prague. The Blackheath design has simple means of construction, a timber armature and a plaster covering; there is however masterly lighting and well proportioned geometry. The other solution exploring step and soffit is a construction made in steel channel and hardwood planks set off with a lean balustrade. The location is commercial hence the abandonment of the 100 mm spacing restriction for the metal rods.

Figure 14.1 'Oakland' Blackheath 1992
a Loft stairs

Figure 14.1b Main stairs, constructed in timber framing and clad with plaster to underside

Stairs, Steps and Ramps 319

c

d

Figure 14.1 Ashland Place, Marylebone, 1992
c and d Steel channel and hardwood risers and treads with matching detail to underside

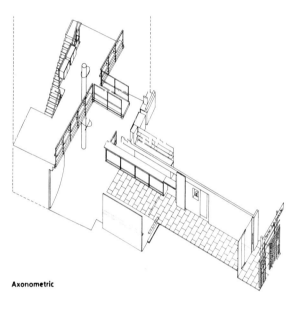

Axonometric

Figure 14.1 Stephen Bull Bistro and Bar, St John Street, Smithfield, 1992
e 10 mm steel plate treads and risers cantilevered from masonry pier

320 Stairs, Steps and Ramps

Case study 14.2 St Annes College, Oxford (1964)

Architects: Howell, Killick, Partridge and Amis

John Partridge was the partner responsible for St Annes College, Oxford completed in 1964. The inclusion of this design from three decades ago is due to sound performance of this straightforward piece of construction. The hardwood strings are propped flight by flight with vertical stability to a balustrade assured by continuous newels. Avoidance of wreathing gives economy. The extension of the rails and string to the 'half tread'

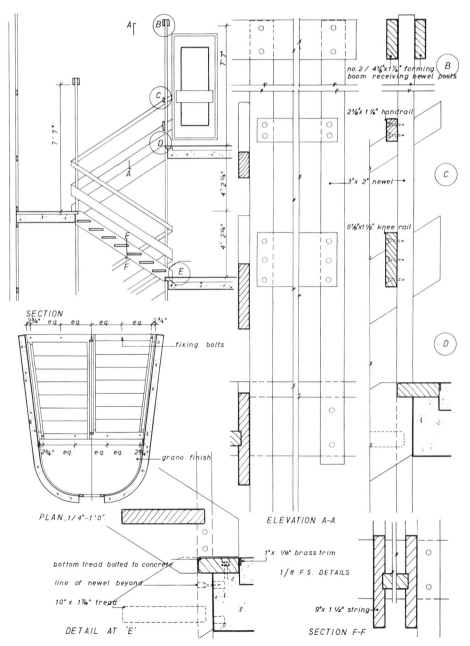

Figure 14.2 St Annes College, Oxford, 1964 a Sheet of details. (From the Architect and Building News)

extension ensures a neat geometry to stair and balustrade at landings. An exemplary essay in fundamentals.

Figure 14.2b General view

Case study 14.3 Tasker Road, Hampstead (1963); 11 Longton Avenue (1977)

Architect: Walter Segal

The zest for staircases has already been mentioned in the introduction with drawings made to full size for appraising the design. Walter Segal's designs for the Tasker Road houses had a compact plan with a compact dog-leg stair supported on a common newel post and a plywood lining at the landing. The assembly was made by George Wade, the joiner who worked with Segal as foreman and contractor in the late 1950s and early 1960s. The carpeted treads had the carpet turned and repositioned over a 20-year period as shown in the detailed sketch. Ideas for a stair that could be screwed together on site from standard battens and planks were developed by the architect and joiner for Segal's own house in Highgate. The designs were further developed with self-build housing work in the 1970s. The illustration showing the stairs made by Ken Adkins in 1977. The cost was a fraction of the traditional kit which incurs case work, newels, balustrades and handrailing.

322 Stairs, Steps and Ramps

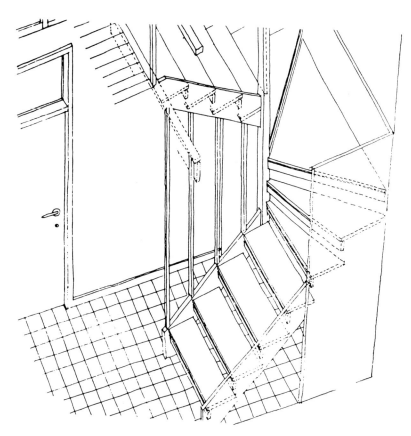

Figure 14.3 Tasker Road Stairs, Hampstead, 1963
a Axonometric

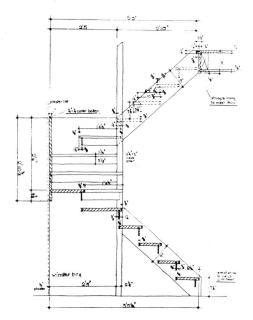

Figure 14.3b General view

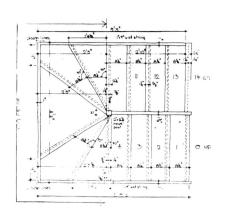

Figure 14.3c Detail of construction

Stairs, Steps and Ramps 323

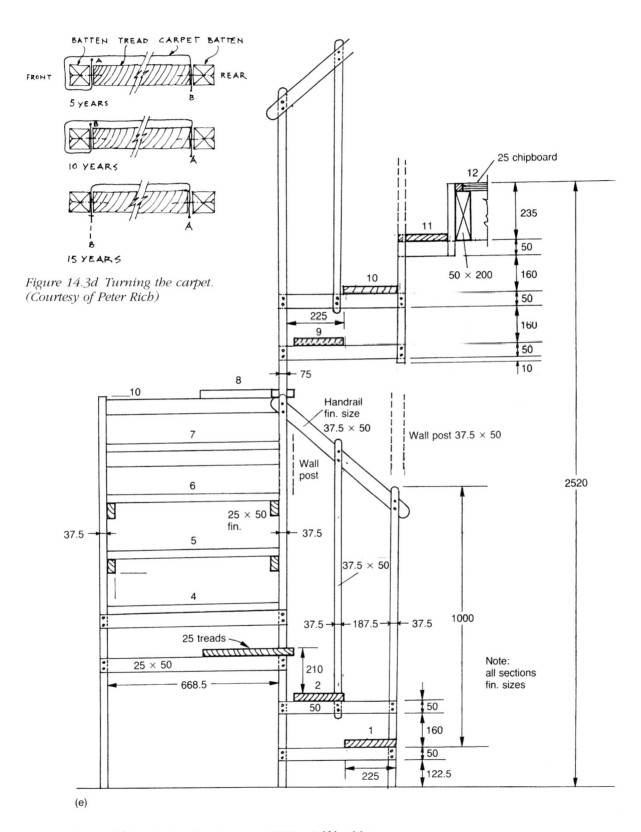

Figure 14.3d Turning the carpet.
(Courtesy of Peter Rich)

(e)

Figure 14.3 11 Longton Avenue, 1977 e Self-build stairs

Metal or steel and glass

Case study 14.4 Lemsford Mill (1985)

Architects: Aldington, Craig and Collinge

The constructional work at Lemsford Mill involved renovation of the structure and the provision of new bridges and stairs with tubular work and forged steel brackets employed for the principal members. The engineering input related to approximate lengths and sizes of connection whilst the architectural detail was involved in the precise outline for connections and profile. The end result is a magnificent piece of detailing in the functional tradition. The project was given the RIBA Regional Award in 1988. The materials used comprise galvanized and painted

Figure 14.4 Lemsford Mill
a External context

plate and tubular steel. The detailer was Paul Collinge with engineering advice by John Austin of Structures and Services Partnership and Mark Whitby.

Figure 14.4c Detail view of external stairs

Figure 14.4b General view of internal stairs

326 Stairs, Steps and Ramps

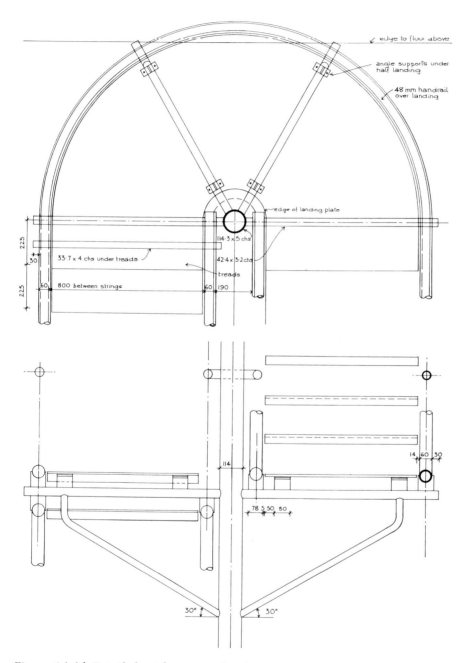

Figure 14.4d Detail sheet for external stairs

Case study 14.5 The Splash Leisure Pool at Sheringham, Norfolk (1989)

Architects: Alsop and Lyall

The swimming baths at Sheringham has an unusual counterbalanced steel ladder to give controlled access to the high diving platforms. The hinged tubular framing can be tilted out of reach in order that the pool and surrounds can be safely used for simple leisure activities. The streamlined detail fits the geometric and weight limitations, the curved flair having an echo of Erich Mendelsohn and Otto Salvisberg. The staircase was first published in the revised Detail Sheets published by the *Architects Journal*. It is significant that stair details have the same frequency of publication as fenestration, in other words, it is one of the more popular aspects in building design.

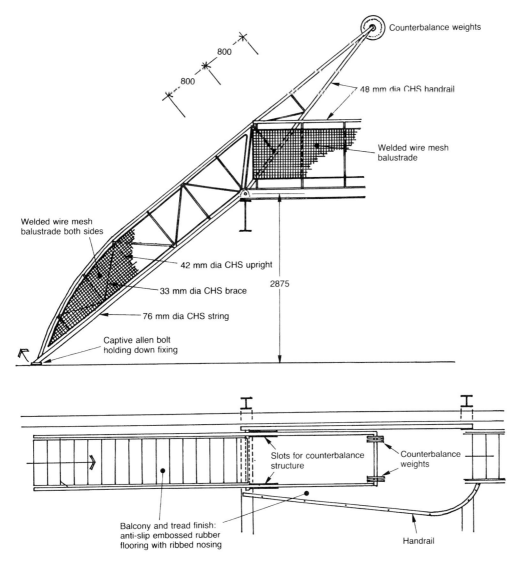

Figure 14.5 Splash Leisure Pool, Sheringham, Norfolk, 1989
a Key elevation and plan

328 Stairs, Steps and Ramps

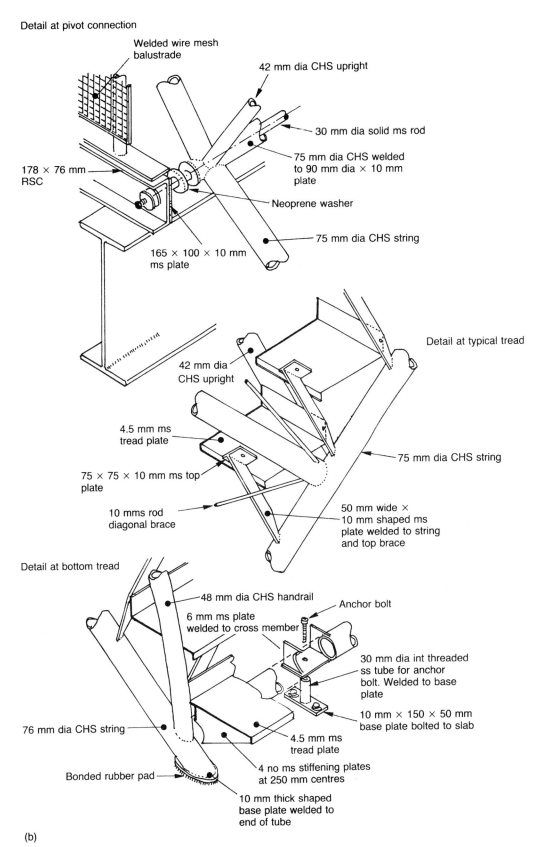

Figure 14.5b Constructional details. (Courtesy of Peter Clash)

Case study 14.6 Office Stair for the 1 Conway Street, London

Architects: Nicholas Grimshaw and Partners

The architects designed, fabricated and erected two staircases for their new offices at 1 Conway Street. The practice had already moved into the premises (formerly a factory) when the kit of aluminium components was designed to be carried into the office, assembled and then manually lifted into position with an absolute minimum of disturbance to the workforce and to the existing structure.

An 'off the shelf' extruded yacht mast, cut to length, forms the stringers with tread supports slotted and locked into the sail groove. Aluminium cast 'fish-heads' are clamped to the stringers to form bearings. At midspan cast 'legs' provide fixing points for rods which strengthen the lightweight frame. These thin steel rods, together with the aluminiun stringers and legs, form trusses which support the stairs. Each stair is designed to meet stringent loading, deflection and vibration criteria.

The kit was produced by various specialist fabricators. The aluminium components were formed from standard sections without welding; the castings were bead-blasted after machining and the rods and rod connectors were supplied by a yacht rigging manufacturer.

The structure was built on trestles in the reception area of the office. The perforated treads were fixed in position one by one. Each stair was lifted into position on ropes and the fish-heads were secured to supporting beams. The handrails were then attached and finally rigged and swagged.

The end result has been successful in both practical and visual terms. The firm is now considering a third stair, in a new location, to be achieved by simply ordering another set of component parts.

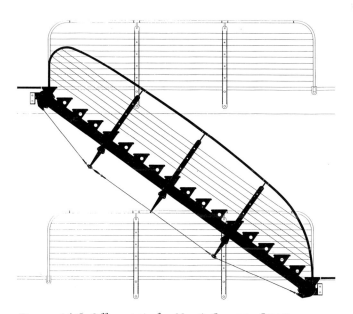

Figure 14.6 Office stair for No. 1 Conway Street
a Outline detail

330 Stairs, Steps and Ramps

Figure 14.6b General view of stairs

Figure 14.6c View of base to stair

Case study 14.7 The Metropolis Recording Studio, London

Architects: Powell-Tuck, Connor and Orefelt

The Metropolis Recording Studio is a remarkable essay in the art of dramatizing stairs. The functional requirements for movement are met within an atrium space which serves all floors. A goods lift rises externally to the curtain wall. Service stairs are close by in the adjacent offices. The very extrovert walkways within the atrium connect to each studio and to the hospitality suite at roof level. A form of stage for the performers, the virtuoso quality is perhaps reflective of the popular music world. The space and the way the atrium is filled is indeed theatrical.

*Figure 14.7 Metropolis Reccording Studio
a General view from ground level*

Stairs, Steps and Ramps 331

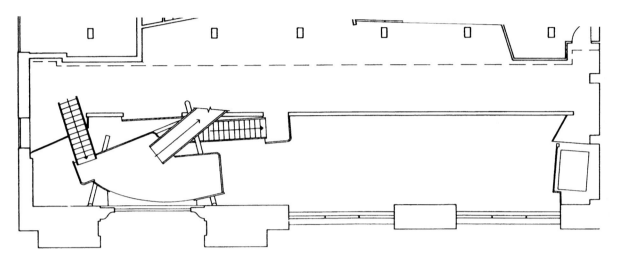

Figure 14.7b General arrangement of upper flight

Figure 14.7c Detail of lift cage

Concrete, stone and clad staircases

Case study 14.8 Ismaili Centre, London

Architects: Casson, Conder

The Ismaili Centre, London has a well-contrived plan which provides for the community's religious, cultural and social needs. The layout is competent in handling large crowds for formal and informal occasions. The sequence of lobbies and stairs that lead from street level to roof garden are of considerable interest. The following explanation has been drawn from Neville Conder's notes.

The main staircases have always been thought of as having special significance and the building has been dubbed 'a staircase to prayers'. The progression to prayers, in fact, comprises two staircases and two extended hallways. First, after the entrance hall, is the inner hall leading the visitor to the far end of the building, before reaching any stair. The reasons are partly to effect a change of mood after leaving the trafficked street and partly to provide, behind columns, a long frontage to cloakroom counters.

On the first floor this distancing by level concourse is repeated again, by walking the visitor along the length of the building: an assistance in preparing for prayers and providing a long frontage for the shoe hall, a significant functional element. The first staircase qualifies as a 'central space stair' while the second stair to the Prayer Hall is a 'periphery' version, although of grander design.

The whole progression is designed to encourage dalliance and to seduce people upwards in gentle stages without too much sense of climb. There are many elderly visitors and as it is not possible to operate lifts that can cope with moving 1 200 people in a few minutes, the landings are crucial – a half landing to the lift hall, lavatories and 'lookout'; another half level to the Social Hall and a further one to reach – finally – the first floor. The Social Hall is indeed a mezzanine, a vast landing without doors to encourage dalliance before prayers and in particular, after prayers. It also assists by extending the exit time for large crowds so that cloakrooms and even the local station are not overwhelmed.

The curvilinear handrailing is designed as a tactile experience: a blind man's stair. The kinked rail in the descents, are a warning, the washers to warn of the first and last riser and the bishops-crook endings that widen in the hand, as if to say farewell.

Figure 14.8 Ismaili Centre, South Kensington Gardens
a Detail of handrailing. (Courtesy of Crispin Bryle)

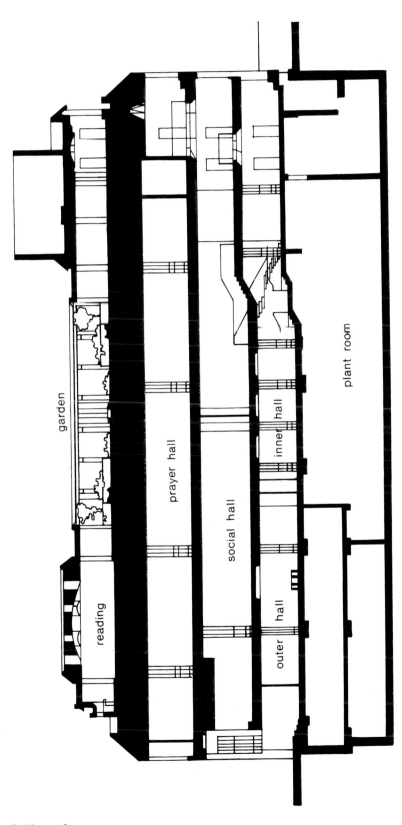

b *Floor plans*

334 Stairs, Steps and Ramps

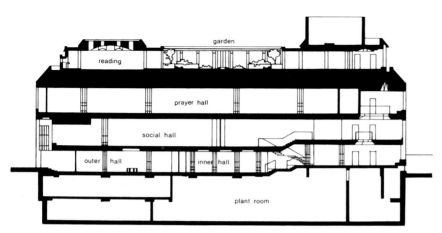

Figure 14.8c Long section

d

e

Figure 14.6d and e Sequential views of stairs from ground floor through to first floor

Figure 14.8f Detail of handrailing

Case study 14.9 Chapel at Fitzwilliam College, Cambridge (1991)

Architects: MacCormac, Jamieson, Pritchard

The new chapel at Fitzwilliam College Cambridge draws on a number of images to create a setting for contemporary worship. The compact design with chapel and crypt placed one over the other within a 15 m drum places considerable emphasis on stairs. There is one stair for the organist and latecomers, and pair of formal entry stairs that symbolically rise from darkness to the light. There are back stairs for the priest and perhaps a means of escape. The formal entry is sculpted as a curving slot between the crypt walls and the enveloping drum. The materials are natural, painted and textured concrete walls and generous oak treads and risers with a 'flashgap' to the walls to overcome cover fillets or the mastic botch.

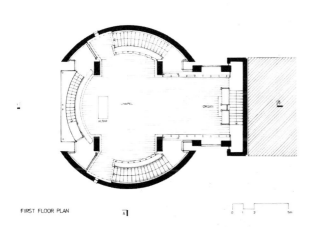

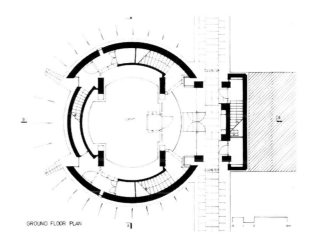

Figure 14.9 Chapel, Fitzwilliam College, Cambridge
a Chapel plan and first floor

Figure 14.9b Crypt plan

336 Stairs, Steps and Ramps

Figure 14.9c Chapel at first floor with main stairs shielded

Figure 14.9d Narrow stairs behind altar

Stone and steel

Case study 14.10 Brasenose College, Oxford (1962); Cripps Building, St Johns College, Cambridge (1967)

Architects: Powell and Moya Partnership

The 1960s found Powell and Moya designing major extensions to colleges in both Oxford and Cambridge. Their designs were distinguished by modern forms of construction but shaped to the scale and cadence of college sets. The materials reflected the context, natural stone for exterior treatment, white concrete with crushed limestone aggregate for framing elements, externally and internally, and traditional oak or stone for the hard used stairs within the buildings. The designs at Brasenose have a sparce simplicity with forged steel shaped from square to circular at the newels coupled to robust oak handrailing and strings. At St Johns, Cambridge, the treads are cantilevered blocks of stone with ironwork equal to Brunel's inventiveness.

Figure 14.10 Brasenose College, Oxford, 1962
a View of landings

b Exterior stone steps and ramp

Case study 14.11 The new British Library, 1995

Architect: Colin St John Wilson

The brief for the entrance hall to the new British Library called for arrangements whereby the major elements of the complex had a discernible presence within the sequence of halls and vestibules. The problem has been solved by using three floors of vestibule in order to make a clear heirarchy in use. Escalators are used in direct ascent level by level together with promenade stairs for the public realm.

The staggering of the building masses in 'U' form permits lifts to be installed in conventional groups that provide alternative

Figure 14.11b View mezzanine level

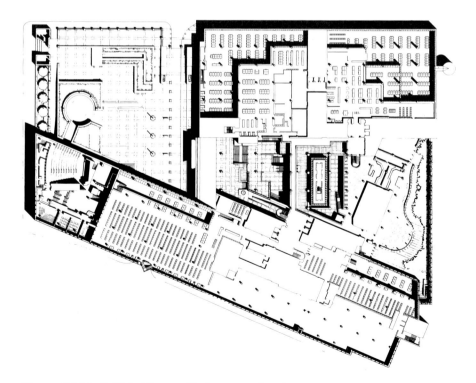

Figure 14.11 British Library, St. Pancras, London
a Layout at entrance and mezzanine level

Stairs, Steps and Ramps 339

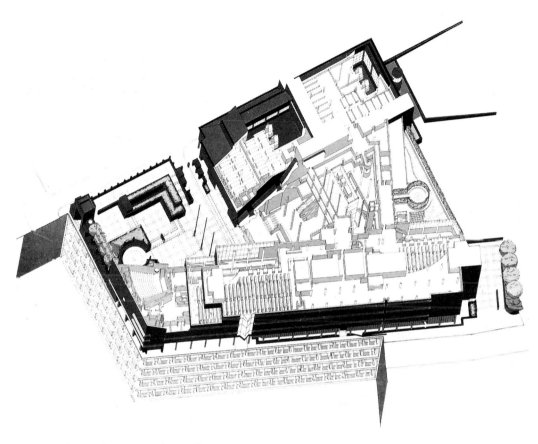

Figure 14.11c Axonometric section

access to the balconies which serve the upper hallways. The top-lit space has generous north lights that brighten the deepest recesses of the layout. The final effect of the layered halls and stairways will create the finest public space constructed in London since the Royal Festival Hall.

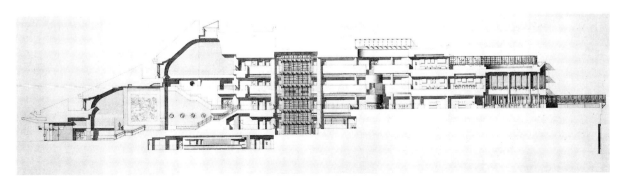

Figure 14.11d Long section through entry hall

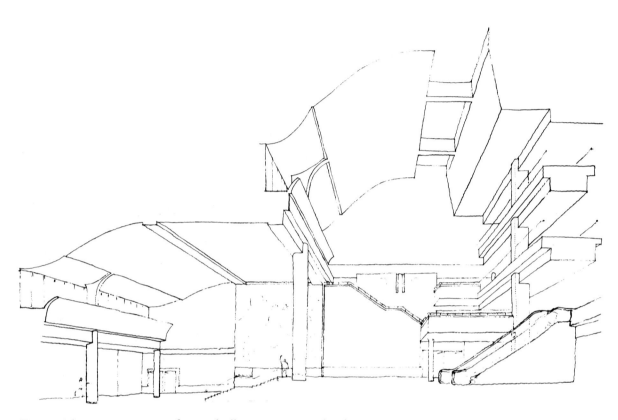

Figure 14.11e Perspective of entry hall at mezzanine level

Case study 14.12 No. 1 Finsbury Avenue, London, for Rosehaugh Greycote Estates Ltd (1985)

Architects and Engineers: Arup Associates

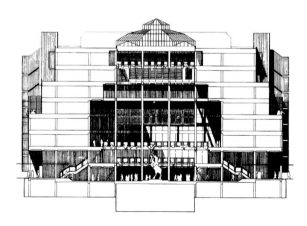

Figure 14.12 No. 1 Finsbury Avenue, London, 1985
a Interior showing stair profile

Significant changes have taken place with office planning in the UK where the changed emphasis contained in new Codes of Practice has been demonstrated and in some cases, anticipated. The seminal layout from the 1980s is the transformation achieved with the deep plan spaces at No. 1 Finsbury Avenue. This remarkable complex set the standard for Broadgate and showed the advantage in placing office areas around atria, with stair and lift cores placed to the periphery. The floor print of the planning modules of 6 m squares expanded to 7.5 or 9 m gave clear working areas of 18 m depth with natural lighting to both sides. The stair nodes articulate the facades into well scaled elements, formed as pavilions set back in their upper storeys to accord with light angles, the staircases forming towers as permitted infringements of set back codes. The geometry of principal and secondary stairs permits a variety of plan arrangements to suit separate tenancies with individual access or the freedom to utilize a 'race track' plan using a triple array of office and service zones.

Figure 14.12c Facade with stair nodes expressed as towers

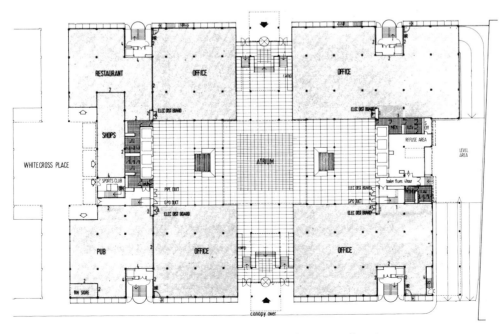

b Footprint at entry level. (See also Figure 4.10e for upper floor)

Case study 14.13 IBM Headquarters, Bedfont Lake, London (1993)

Architects: Michael Hopkins and Partners

The atrium concept is expanded for the IBM headquarters, Bedfont Lake to embrace a central working area. The principal lifts and vertical circulation spring from the bridges spanning the entry axis and connecting to continuous balconies at all levels. The centralized volume could be compared to the Forum of Trajan with upper walkways, and various 'streets' which unite all the working areas. The transparency given by current glazing techniques enables secondary stairs, formerly closed away in hidden cores, to be revealed to view and to fulfil the visual role accorded historically to stairs within buildings.

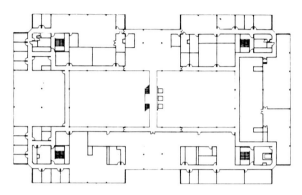

Figure 14.13 IBM Headquarters, Bedfont Lake, London, 1993
Figure a Upper floor plan

Figure 14.13c View of secondary stairs

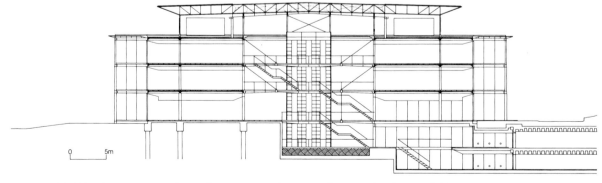

Figure 14.13b Cross section through atrium with view of access bridge and lifts

Figure 14.13d Detail of metalwork

Case study 14.14 Apple Computers facility, Stockley Park (Phase 1 1989, Phase 2 1992)

Architects: Troughton McAslan

The final example is modest in scale but demonstrates the way central space stairs can satisfy planning criteria within low rise buildings. The Apple computer offices by Troughton and McAslam at Stockley Park fulfill the fixed constraints with standardized framing grids and storey heights. The typology aimed at 18 m depth office or working areas having natural light, with a common access or service core of 9 m wide to absorb foyers, vertical cores and services. The units could be let in simple ownership or as multiple tenancies with separation of entry stairs as required. External escapes could also be added where necessary.

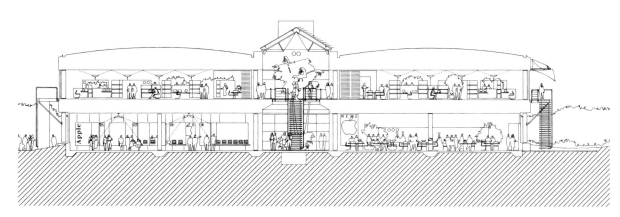

Figure 14.14 Apple Computers facility, Stockley Park
a Cross section

344 Stairs, Steps and Ramps

Figure 14.14b General view

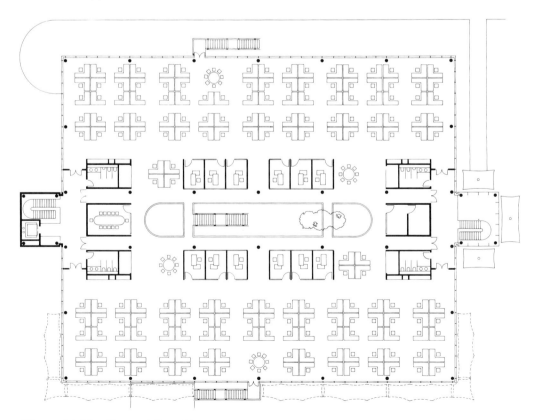

Figure 14.14c Ground floor plan

Case study 14.15 No. 3 Minster Court, City of London (1993)

Architects and engineers: YRM plc

Architects to core building: GMW

This concluding case study refers to the role today of escalators in providing drama and movement through large vertical spaces. A prime example of this direction is the atrium to Minster Court which serves 300,000 square feet at the London Underwriting Centre. The trading pattern is similar to Lloyds (see Figure 12.5j, k, and o) the brokers numbering 1500 at any one time for the operation of Minster Court. This volume of traffic between floors is handled by a bank of 16 escalators stacked in scissor formation across the centre of the atrium. There are four lifts in each of the two office cores to service longer journeys and to speed staff movement in the mornings and at the close of day.

The circular atrium with its movement systems is truly the heart of the complex and provides an exemplary example of escalator design. The grouping is the tallest assembly completed to date world wide and is arranged with suspension from four vertical steel rods. These open into fan shapes to connect with a metre deep ring beam at roof level. This in turn is carried by 12 pairs of columns within the atrium structure. This may appear elaborate but a fire during the construction

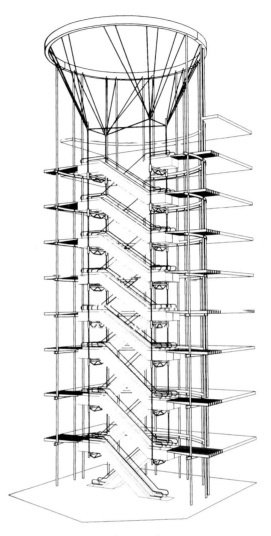

Figure 14.15 London Underwriting Centre, No. 3 Minster Court, City of London, 1993
a Perspective of escalators, suspension system and atrium. (YRM plc)

phase meant that some rebuilding was needed. The client and design team for the fitting out decided to insert escalators instead of wall climber lifts in the original concept. The end result is one of the most inviting entry halls built in the City of London.

346 Stairs, Steps and Ramps

Figure 14.15b Interior view looking upward to suspended escalators within atrium. (Courtesy of Peter Cook)

Epilogue

Traditionalists will complain that mechanisms can never replace the sculptural possibilities in the art of staircase construction with curving and/or diagonal alignment. Further complaints will follow that noisome intrusions of machinery are no replacement for the static, silence and evocative delights of an empty flight of steps. Bernini's Scala regia at the Vatican needs no human touch.

There are other circumstances where conventional stairs provide not only a visual centrepiece but also a meeting point in public places. The designs by Garnier at the Paris Opera have already been mentioned in the Introduction and it seems pertinent to close with a similar concept that forms the core of Richard Horden's design at Epsom. Twin flights are employed to accommodate the crowds but they also provide that essential visual delight of stairs in public places; they provide a platform to view the interior but equally important they provide a stage to be seen from. A trim modern-day version of Garnier's Parisian masterpiece.

Figure 14.16 View towards double curved stairs at the core of the Queens Stand, Epsom Racecourse, 1993. (Richard Horden) (Courtesy of Spiral Staircase Systems)

Index

Aalto, A., 1, 64–65, 222
Aarhus, City Hall, 102, 105–6, 206, 274
Abbey of Mont Saint Michel, 259
Accommodation stairs, 91–96
Acropolis, 16
Adam, R., 207
Adams, Park, Colombo, 130
Adams, W.H., 158
Adkins, K., 321
Adler & Sullivan, 181
Ahrends, Burton and Koralek, 92
Aidrail Ltd, 203
Alberti, 10, 11
Albini, F., 190, 194
Aldinton, Craig & Collinge, 324–326
Alexander the Great, 130
Alhambra, 16
Allies & Morrison, 317
Alsop and Lyall, 327
American codes, 252–258
Amersfort Art Gallery, 109
Apple Computers, Stockley Park, 343–344
Architect & Building News, 1
Architectural Review, 32
Arneberg, A., 107
Arup Associates, 79, 186, 340–342
Ashland Place, Marylebone, 317–319
Asplund, G., 1, 3, 102, 117–118, 191, 272–273, 310
Assyria, 7
Astaire, F., 30
Atrium, 74
Augustusburg, Bruhl, 100–1
Aukett, M., 288

Baker House Dormitory, 64–65
Balustrade and handrail, 169–177, 195–208, 215, 217, 221, 226–227, 241–244
Basel University, 291
Basilica, Vicenzia, 114, 116–117
Beauvais Cathedral, 259
B.D.P., 33
Bebb, M., 83

Belcher & Joass, 88
Bellahoj, Copenhagen, 65
Behnisch, G., 99, 294–295, 306
Benn Levy House, Chelsea, 43
Bens A. and Kriz J, 88, 90
Berkeley House, 269
Bergoglio & Mutti, 217
Bernini, G.L., 21, 347
Biblioteca Laurenziana, Florence, 11
Bistro and Bar, St John Street, 319
Blanc, L.D., 268–269, 290
Blois, 25, 228
Bofil, R., 64
Bohm, G., 64, 103, 108, 184
Borgias, 132
Boris, G., 32
Boston City Hall, 106, 108
Botrea Stairs, 180
Boyle, R., 20
Bracken House, 80
Bradbury Building, 80, 92, 96, 264, 272
Brasenose College, Oxford, 337
Breuer, M., 66, 190–191, 194, 293, 307
British Library, 338–340
British Museum, 114, 118–120
British Pavilion, Seville Expo, 288–290
Britz Siedlung, Berlin, 65
Bruckland, J., 180
Bruegal, 7
BSIF Prefabricated House, 142
Buchanan, P., 281
Building codes, 44, 68–69
Bull, S., 317, 319
Burlington House, 114, 118–119
Burnet, J., 120
Burnham, J., 275
Burton, D., 262

Cafe Astor, 88, 90
Cantilever stairs, 229
Capitol, Rome, 21, 23, 151
Capitoline Museum, 21
Carcassone, 221

Carey, J., 142
Carita Monastery, Venice, 228, 231
Carpenters Assistant, 27, 165, 170, 180
Carter, B., 1
Casa Batllo, 28
Casa Mila, 29
Caserta, 147, 149–150
Casson, Conder, 234–334
Cast iron and steel stairs, 181
Centraal Beheer, Apeldoorn, 309
Centre Pompidou, 95, 184, 280–282, 290, 304
Centrosoyus Building, Moscow, 121, 125, 210
Chaldean ramped temples, 7
Chamberlain, P., 195, 200
Chambord, 19, 24–25, 228
Chamundi, Mysore, 130
Chapel at Fitzwilliam College, 335–336
Charles de Gaulle Airport, 288
Chartered Institution of Building Service Engineers, 259
Chatsworth House, 147, 151, 231, 233
Chermayeff, S., 46, 292
Chinese Garden, 158
Chiswick House, 19, 228
City Hall, Arnhem, 192
Civic and public stairs, 98–129
Clare Hall, Cambridge, 215–216
Clore Gallery, 112, 115
Coates, W., 64, 66
Codes of Practice, 195, 199, 208, 236–258
Collymore, P., 45
Colosseum, Rome, 8, 9, 209
Colosseum Theatre, London, 259, 262
Commercial stairs, 68
Communication Tower, Tibidabo, 296–297
Concrete stairs, 209–227
Conservatori, Rome, 21
Courthouse Gothenburg, 3, 114, 117–118, 191, 272–273

Covent Garden, Opera, 33, 107
Crabtree, W., 97
Crane, H.B., 269
Cripps Building, St Johns College, 337
Criterion, 70, 73
Crown Reach apartments, 64–65
Crystal Palace, 87–88
Cullen, G., 153, 155
Cylindrical and spiral stairs, 23–29

Da Costa House, Highgate, 233–235
Darborne & Darke, 61–62
da Vinci, L., 4, 19–20, 24–25
De La Warr Pavilion, Bexhill, 292
De Mille, C., 270
De Sanctis, F., 135
Decorative Arts Museum, Frankfurt am Main, 301–303
Designs in Exile, 97
D.H. Evans, London, 269
Direct flights, 12–16, 39–43
Dixon, J., 33, 107
Doldertal, Zurich, 66
Dollman, G., 259, 261
Dominia, A.S., 65
Domitian, 8
Dog-legs, 43–44
Dom-Ino House, 49–50
Dreir, H., 30
Duchess de Chatearoux, 259

Eaton Center, Toronto, 314–316
Economist Building, 80–81, 83
Edoux, 261, 263
Edwardian Garden, 158
Eiffel Tower, 181, 261, 263, 264, 277, 280, 290
Einstein Village, Princeton, 194
El Castillo Chihen-Itza Yucatan, 8
Electricity Board, Prague, 88, 90
Elevators, 259–267
Emberton, J., 1
Epidauros, 29–30
Erickson, A., 144
Erskine, R., 215–216
Escalators, 85, 267–272
Escape in case of fire, 242–252
European Investment Bank, Luxembourg, 110–114, 129
Evans, Sir A., 17
Exhibition House, New York, 293
Express Lift Co., 281
External stairs, 130–158

Falling Water, 292
Familistere, Guise, 58–59
Farrell, T., 54, 107, 109

Federal Offices, Seattle, 145
Federation du Batiment, Paris, 193
Fasenenplatz, Berlin, 64
Financial Times, 78, 80
Finsbury Avenue, No. 1, 76, 79, 340–341
Fire Brigade Requirements, 59, 61, 63, 69, 80–81, 247–252
Fire escape, 182–184
Fitzroy Robinson, 87
Fletcher, Sir B., 12
Foster, N., 79, 119, 186, 193, 203, 272–273, 277–281, 290, 296–297
Fourier, C., 58
French Garden, 158
Fry, M., 43, 169

Galeries Lafayette, 88–89
Gamble House, Pasadena, 160, 176
Garnier, C., 3–5, 30–32, 107, 270, 347
Gatwick Airport, 287
Gatwick Hotel, 82, 86
Gaudi, A., 23–24, 28–29, 45–46, 51
Gavois, J., 259–263, 290
Generalife Gardens, Granada, 133
Georgetown Mall, 91
Gerschaftshaus, Vienna, 95
Gibberd, F., 136, 142
Gibbs, J., 123
Gimbel's store, Philadelphia, 268
GMW Architects, 343–345
Goetch, K., 259
Goetheanum, Dornach, 211, 217
Going, 236–240
Goldfinger, E., 49, 51, 210
Gone with the Wind, 30
Goodhart-Rendel, 32
Greene and Greene, 10, 160, 176
Great Wall of China, 143
Grillet, 147, 151
Grimshaw, N., 182, 288–290, 329–330
Gropius, W., 43, 169
Guggenheim Museum, 26–27, 33, 210
Guinard, H., 88–89
GUM store, Moscow, 88

Haag, R., 145
Hallidie Building, 92, 97, 182
Halprin, L., 153, 155, 158
Hammond, Beeby & Babka, 271
Hampton Court, 233
Harold Washington Public Library, 271
Harrods, London, 268–269, 290
Harvey Court, Cambridge, 64, 66
Havas Conseil, Paris, 207

Hatfield House, 162, 165
Heals, London, 275, 290
Hentrich, H. and Petschnigg, H., 78
Heriots Hospital, 57
Herron, P., 274
Hertzberger, H., 309
Hewi (UK) Ltd., 170, 208
Highbury Terrace Mews, 45
Hille, 194, 198
Holland, J., 195, 200
Hollein, H., 95
Holy Ghats at Benares, 131–132
Hong Kong Club, 307–309
Hongkong Shanghai Bank, 4, 78–79, 277, 279–281, 287,
Hooke, R., 231–232
Hopkins, M. and P., 51, 53, 78, 80, 272, 342
Horden, R., 347
Horta Museum, 28
Horta, V., 28–29, 176, 180, 186
House at Blackheath, 317–318
Howe, G., 269
Howe, J., 43
Howell, Killick, Partridge and Amis, 320
Hradcany Castle, Prague, 126, 129
Hyatt Regency, Detroit, 84–85
Hydraulic lifts, 265

IBM, Bedfont Lake, 76, 113, 342–343
Ibn Tulun, Cairo, 8, 160, 172
Institute du Monde Arabe, Paris, 274–275
Isfahan, 7
Ismaili Centre, London, 332–334
Italian Gardens, 158

Jacobs Ladder, 9, 10
Jacobsen, A., 102, 105–6, 183, 188, 206, 272, 274, 291, 298–299
Jahn, H., 300–301
Jeanneret, P., 49–50
Jellicoe, G.A., 137, 153, 158
Jencks, C., 51, 54
Jiricna, E., 1, 74, 91, 94, 97, 182, 191, 194–195
John Carr Joinery Sales Ltd, 180
John Hancock Centre, Chicago, 78, 266
John Lewis Stores, 89–90, 92, 97
John Soane Museum, London, 206
Johnson Building, Racine, 272
Jones, I., 228, 232
Josephs store, 91, 94–95, 194, 199–200
Jourdain, F., 88, 186

Kafka, F., 8
Kallmann, McKinnell & Knowles, 108
Keddleston Hall, 207
Kempster, W., 233
Kennedy Memorial, Runnymede, 130, 150, 153–154
Kenny, S., 269
Kensington Traders Ltd, 208
Keswick, M., 158
Kohn Pederson Fox & Smith, 270
Kramer, F., 52, 55
Kremlin Palace of Congresses, 269
Kunsthistorisches Museum, Vienna, 102–3, 119

Lacey and Jobst, 64
Ladders and steps, 52–55, 160–162, 181–182
Landscape of Man, 158
Lanesborough Hotel, 83, 87
Larkin Building, 287
Lasdun, D., 3, 32, 70, 74, 110–111, 129, 211, 217
Law Courts, Vancouver, 139, 144
Lawn Road Flats, 66
Layout plans, 13–14, 138–139, 141, 145, 148
Le Corbusier, 3, 49–50, 58–59, 121–125, 188, 193, 210, 302
Le Grand Arche, Paris, 275–277
Le Grand Louvre, Paris, 304–305
Le Notre, 152
League of Nations Competition, 121, 125, 210
Legends Club, 74, 182, 191
Leisure Pool, Sherringham, 182, 327–328
Lemsford Mill, 324–326
Les Espaces d'Araxes, 64
Lescaze, W., 269
Lifts, 85
Lift cores, 56–59, 81
Lillington Gardens, 61–62
Linderhof, 259, 261
Lindisfarne Castle, 147, 149
Lloyds HQ, 4, 69, 78–79, 270, 280–281, 285–287, 290
Loebl, Achlossman, Bennett and Partners, 270
London Museum, 275
London Theatre, 269
London Underground, 267–268
Longton Avenue, 321, 323
Loos, A., 317
Lovejoy Plaza, 153, 155–156
Louis XV, 259
Louvre Pyramid, 186
Lubetkin, Drake & Tecton, 291
Ludwig II of Bavaria, 259, 261

Luisa Parisi, 176, 180
Luther, M., 21
Lutyens, E., 134, 147, 149, 158

McGrath, R., 46–47
MacCormac, Jamieson, Pritchard, 335–336
MacFarlane, D., 182, 191
Maison Carrée, Nîmes, 16–17, 114, 118–119
Maison Citrohan, 49
Maison Clarte, 63, 188, 193
Maison de Verre, 186
Majorelle, L., 88–89
Mallet-Stevens, R., 30
Manchester Airport, 288
Manser, M., 85
Marco Polo, 130
Marquise de Pompadour, 259
Martin, L., 64, 66, 76, 127–128
Masson, G., 158
Matthew, R., 127–128
Maufe, E., 1, 275, 290
Mausoleum of Aga Khan, Aswan, 145
Meier, R., 301–303
Memorial de la Deportation, Paris, 131–132
Mendelsohn, E., 1, 46–47, 88, 90, 97, 292, 327
Menzies, W.C., 30
Metropolis Recording Studio, 330–331
Mewes & Davis, 83, 87
Mexico, 7
Michelangelo, 11, 21, 23
Michelozzi, M., 137
Milton Gate, 70, 74
Minster Court, City of London, 345–346
Moller, E., 102, 105
Monument, City of London, 231–232
Moro, P., 194, 198
Muller House, Prague, 317
Multi-turn flights, 16–20

National Building Regulations, 42, 52, 56, 62–63, 68, 75, 82, 97, 135, 169–170, 200, 236–258
National Gallery, London, 21, 114, 118–119, 121
National History Gallery, 274
National Theatre, London, 3, 29, 32, 110–111, 129, 211, 217
Nelson, P., 30
Nervi, P.L., 209–210
Nestle Headquarters, Vevey, 123, 126–127

Neue National Galerie, Berlin, 201, 208
Neue Pinakothek, Munich, 114, 119, 122
Neutra, R., 10, 292
New York Crystal Palace Exhibition, 261
Newmann, J.B., 101–102
Nikè Apterous, 16, 228
Norton Engineering Alloys Co. Ltd, 201, 208
Nouvel, J., 275
Novello & Lange, 211, 217
Novo Factory, Copenhagen, 188

Office cores, 71–78
Offices at No.1 Conway Street, 329–330
O'Hara Airport, Chicago, 288
Ontario Place, Toronto, 314
Oslo City Hall, 103, 107
Ostia, 56–7
Otis, E.G., 261
Otis Elevator plc., 259–264, 277, 290
Ottewill, D., 158
Ove Arup & Partners, 291, 296–297
Our Lady of Light Chapel, Sante Fe, 160, 176
Oxford Physic Garden, 231

Paimio Sanitorium, 222
Palace of Knossus, 17–18
Palace of Soviets Competition, 121, 126
Palace of Westminster, 98, 100
Palais de Justice, Brussels, 8, 102–5
Palais Royal, 87
Palazzo Capra, 228, 231
Palazzo Grimani, 112, 114, 117
Palazzo Iseppo Porto, 116
Palazzo Municipio, Genoa, 18, 105, 228
Palazzo Senatorio, 21
Palladio, 29, 114, 117, 228, 231, 233
Palm House, Kew, 181, 185
Paris International Exhibition, 261, 264, 268
Paris Metro, 88–89
Paris Opera, 3–5, 30–32, 107, 347
Paris Opera (new), 29
Parliament Building, Bonn, 98–99
Parthenon, 16
Paternoster lifts, 265
Pawley, M., 97
Paxton, J., 87–88
Peabody Trust, 56
Pei, Cobb, Freed and Partners, 304–305
Penguin Pool, London Zoo, 291

Perret, A., 209
Philadelphia Savings Fund Society, 269
Phönix-Rheinrohr AG, 77–78
Piano, R., 278, 282
Pick, F., 268, 290
Piranesi, 4, 6, 8, 84
Plecnik, J., 126, 129
Polk, W., 97
Polaert, 103
Polycleitos, 29
Ponti, G., 294
Portman, J., 82–83, 97
Portmerion, 151, 154–155
Poulsson, M., 107
Powell and Moya, 337
Powell, Smith & Billington, 97
Powell-Tuck, Connor & Orefelt, 330–331
Powys Castle, 140
Price and Myers, 317
Price, S., 234
Propylaea, 15, 118, 228

Queen Anne's Gate, 166
Queen Anne's Mansions, 59
Queen's House, Greenwich, 231–233
Queens Stand, Epsom Racecourse, 347
Quincy Market, 106

RAC Building, Milan, 294
Randall, P., 62
Ranzenberger, H., 217
Reilly, C., 97
Reitveld-Schroder House, 42–43, 109
Reitveld, G., 110
Renaissance Center, Detroit, 81, 83
Renishaw, 134, 140
Renton Howard Wood Levin, 70, 73
Research & Development Group, MOHLG, 62–63
Research House, Silverlake, 292
Residenz, Wurzburg, 101–102
Rheinberg Town Hall, 105, 108, 184
RIBA HQ, 142
Rice, P., 280–284
Richard Burbridge & Sons Ltd, 180
Rise, 236–240
Ritz Hotel, 83, 87
Roche Chemicals, Welwyn Garden City, 165, 167–168
Rochester Castle, 181, 183
Rødovre Town Hall, Copenhagen, 298–299
Rogers, R., 50, 52, 69, 79, 272, 277–278, 282–287, 290
Rohn, R., 3, 291
Roth Brothers, 66

Roux-Combaluzier, 262, 264
Royal College of Physicians, 110, 114, 129
Royal Festival Hall, 114, 123, 127–128, 338

Sackler Museum, Harvard, 112, 115
Sagrada Familia, 23–24, 28, 228
Sainsbury Arts Centre, Norwich, 186, 193, 273–274
Sainsbury's Store, Camden Town, 288–289
St Annes College, Oxford, 320–321
St Annes Hill, Chertsey, 46–48
St Barnard Monastery, Paris, 25
St Martins in the Fields, 123
St Pauls Cathedral, 233
St Peters, Rome, 21, 23
Salvisberg, O., 1, 3, 165, 167, 183, 187, 327
Samaritaine Store, 88, 186
San Souci, Potsdam, 147, 152
SAS Headquarters, Stockholm, 309–313
Sauvage, H., 88
Scala Santa, 21
Scharoun, H., 306
Schatner, K.J., 119, 124
Schindler, 10
Schinkel, 85, 88
Schocken Stores, 88
School of Journalists, Eichstatt, 121, 124
Schuster, F., 1, 3, 10, 101, 167, 177, 187, 221
Segal, W., 1, 42, 55–56, 167, 321–323
Seidler, H., 306–309
Selfridges, London, 275
Semper, G., 102, 270
Serlio, 57
Shepherd, J.C., 137, 158
Shepherd, P., 158
Sherringham Swimming Pool, 182, 327–328
Shrublands, Chalfont St Giles, 46
Siemens, 261
Scissors lifts, 265
Sitwell, G., 134, 140
Slater and Moberley, 97
Smirke, R., 120
Smithson, A. and P., 83
S.O.M., 83
South Waker Drive, Chicago, 270
Spanish Gardens, 132–133, 158
Spanish Steps, Rome, 133, 135, 151
Speer, A., 8
Spiral stairs, 44–49, 184–186, 215–218, 229
Staatsbibliothek Preussischer Kulturbesitz, Berlin, 306–307

Stamford Assembly Rooms, 160
State of Illinois Centre, Chicago, 300–301
Steiner, R., 211
Stirling and Wilford, 112, 115
Stirling Hotel, London Airport, 82, 85
Stock Exchange Building, Chicago, 181
Stockholm Exhibition, 310
Stockley Park, 184, 186, 188
Stone, N., 231, 233
Stonework stairs, 228

Talman, W., 233
Tapered flights, 21–23
Tasker Road, Hampstead, 321–323
Tate Gallery, 114, 118–119, 122
Taut, B., 64–65
Taylor, R., 233
Temple of Ammon, Der-el-Bahari, 13
Temple at Dendera, 13, 15, 237
Teatro Olimpico, Vicenza, 29
Temple at Chidambaram, India, 130
Tijou, 233
Timber stairs, 159–180
Tiranti, 1
The Architect as Developer, 97
Theatrical stairs, 29–35
Torp, N., 309–313
Tosca, 32
Tower of Babel, 7
Trafalgar Square, 132
Travelatorscape, 287–290
Treads, 162–163, 173, 179, 184, 189–190, 219–223, 232–235
Treppen, 1, 3, 10, 101, 167–169, 177, 187, 198
Troughton McAslan, 188, 343
Trump Tower, 91, 94
Tschumi, J., 126
Turner & Burton, 181
TV AM, 107, 109
Twist & Whitley, 216

UNESCO Centre, Paris, 123
Unite d'Habitation, Marseilles, 58
Universal Studios, 30
University Library, Eichstatt, 294–295

van der Rohe, M., 201, 208
van Gough Museum, 110
Vanvielli, 149
Vasari, G., 11
Vatican, 21, 26
Venturi, R., 21
Versailles, 147, 149–150, 152-3, 259
Via Garibaldi, 20
Villa at Bullono, 217

Villa Capra, Rotunda, 19, 116, 217
Villa d'Este, 132–135
Villa Farnese, 231
Villa Garzoni, 134, 137
Villa Godi, 116
Villa La Roche-Janneret, 3, 49
Villa Medici, Fiesole, 134, 137
Villa Savoye, 49, 210, 301
Villa Thiene, 116
Vitruvius, 10
von Knobelsdorff, C.W., 152
von Spreckelsen, J.O., 276

Wade, G., 321
Wall climber lifts, 74, 82

Water Tower Place, 91, 93, 270
Waterloo Terminal, 182
Wates Housing Development, 218
Weaver, L., 158
Weissenhof, 59
Westminster Cathedral, 260
Whitby and Bird, 317
Wilkins, W., 121
Williams, O., 1
Williams-Ellis, C., 151, 154
Willis Faber & Dumas Building, 277–278, 287
Wilson, C. St J., 66, 338–339
Wingspread, Racine, 292
Winslow House, 56
Winter, J., 181

Wisniewski, E., 306
Wheeler, B., 259
Whiteleys Store, 88
Wood House, Shipbourne, 169
Wren, C., 68, 231–233
Wright, F.L., 10, 26–27, 56, 140, 147, 210, 264, 272, 287, 292
Wyman, G.H., 92, 96, 264

YRM Architects, 86, 287, 343–345

Zeidler Roberts Partnership, 314–316
Zublin-Haus, 112, 116